ALSO BY CHRIS STRINGER

Homo britannicus

The Complete World of Human Evolution
(with Peter Andrews)

African Exodus
(with Robin McKie)

In Search of the Neanderthals
(with Clive Gamble)

LONE
SURVIVORS

LONE
SURVIVORS

HOW WE CAME TO BE
THE ONLY HUMANS
ON EARTH

CHRIS STRINGER

Times Books
Henry Holt and Company
New York

Times Books

Henry Holt and Company, LLC

Publishers since 1866

175 Fifth Avenue

New York, New York 10010

www.henryholt.com

Henry Holt® is a registered trademark of Henry Holt and Company, LLC.

Originally published in the United Kingdom in 2011 by Allen Lane as *The Origin of Our Species*

Library of Congress Cataloging-in-Publication Data

Stringer, Chris, 1947–

    Lone survivors : how we came to be the only humans on earth / Chris Stringer.

        p. cm.

    Includes index.

    ISBN 978-0-8050-8891-5

  1. Human beings—Origin.   2. Human evolution.   I. Title.

    GN281.L65 2012

    599.93'8—dc23   2011030434

First U.S. Edition 2012

Designed by Meryl Sussman Levavi

1   3   5   7   9   10   8   6   4   2

To the memory of lost family Tony and David,
and lost colleagues Bill, Clark, and Roger

Contents

Illustrations

LONE
SURVIVORS

Introduction

WE HAVE JUST CELEBRATED THE 150TH ANNIVERSARY OF THE publication of Charles Darwin's *On the Origin of Species* and his two hundredth birthday, and evolution by natural selection is now widely accepted. But what do we know about the origin of our own species, *Homo sapiens*? Despite the fascinating and growing record of very ancient prehuman fossils, one topic has dominated recent scientific and popular discussion about evolution: our own origins. While it is generally agreed that Africa was the homeland of our earliest human ancestors, a fierce debate continues about whether it was also the ultimate place of origin of our own species, and of everything that we consider typical of our species, such as language, art, and complex technology. Originally centered on the fossil record, the debate has grown to encompass archaeological and genetic data, and the latter have become increasingly significant, now even including DNA from Neanderthal fossils. Yet much of these new data and the discussions surrounding them are buried in highly technical presentations, scattered in specialist journals and books, so it is difficult for a general readership, however informed, to get an accessible overview.

In this book I want to try and provide a comprehensive—but comprehensible—account of the origin of our species from my position in these debates over the last thirty years or so. I've worked at the Natural History Museum in London even longer than that, and the idea that I could have ended up there, studying our origins, was a boyhood dream which I never thought would actually come to pass, given my relatively humble origins in the cockney area of east London. But with supportive parents and foster parents and some teachers who encouraged me along the way, I started to realize that dream when, at age eighteen, I made a last-minute switch from studying medicine to taking a degree in anthropology. It was a gamble that paid off when I was accepted into the Ph.D. program at Bristol University in 1970 to study my favorite fossil people—the Neanderthals—and then even that was capped by the offer of a job in the Palaeontology Department at the Natural History Museum in London, in 1973.

It has been such an exciting time to be working in this field, with wonderful new fossil finds, but also the arrival of a host of new techniques to date and study them. I hope my book will make every reader think about what it means to be human, and change his or her perceptions about our origins—writing it has certainly changed some of mine!

I regularly give talks on human evolution and receive hundreds of inquiries on this topic every year from the media and the public. The same questions recur time and again, and in this book I will try to answer them. These questions include:

1. What are the big questions in the debate about our origins?
2. How can we define modern humans, and how can we recognize our beginnings in the fossil and archaeological record?
3. How can we accurately date fossils, including ones beyond the range of radiocarbon dating?
4. What do the genetic data really tell us about our origins, and were our origins solely in Africa?
5. Are modern humans a distinct species from ancient people such as the Neanderthals?
6. How can we recognize modern humans behaviorally, and were traits such as complex language and art unique to modern humans?
7. What contact did our ancestors have with people like the Neanderthals, and were we the cause of their extinction?
8. Do archaic features in modern human fossils and genes outside Africa indicate hybridization?

9. What does DNA tell us about the Neanderthals and possible inter-breeding with modern humans?
10. What can we learn from a complete Neanderthal genome, and will we ever clone a Neanderthal?
11. What forces shaped the origins of modern humans—were they climatic, dietary, social, or even volcanic?
12. What drove the dispersals of modern humans from Africa, and how did our species spread over the globe?
13. How did regional ("racial") features evolve, and how significant are they?
14. What was the "Hobbit" of the island of Flores, and how was it related to us?
15. Has human evolution stopped, or are we still evolving?
16. What can we expect from future research on our origins?

It is now over twenty years since the publication of the seminal *Nature* paper "Mitochondrial DNA and Human Evolution" by Rebecca Cann, Mark Stoneking, and Allan Wilson that put modern human origins and "Mitochondrial Eve" on the front pages of newspapers and journals all over the world for the first time. Not only did that paper focus attention on the evolution of our own species, but it also led to a fundamental reformulation of scientific arguments about the way that we look at our own origins. A year after that publication, I wrote the paper "Genetic and Fossil Evidence for the Origin of Modern Humans" for the journal *Science* with my colleague Peter Andrews, which set out the contrasting models of modern human origins that have dominated debate ever since: the Recent African Origin model and the Multiregional Evolution model. Later in the book we will see how these models have fared in the face of many new discoveries, but in the first chapter I will look at some of the big questions of modern human origins, including what diagnoses our species, what the recent debates are all about, and how the different models lay out expectations of what we should find in the record of modern human evolution, from fossils, archaeology, and genetics.

1

The Big Questions

IT IS BARELY 150 YEARS SINCE CHARLES DARWIN AND ALFRED Russel Wallace presented their ideas on evolution to the world. A year later, in 1859, Darwin was to publish one of the most famous of all books, *On the Origin of Species*. Then, the first fossil human finds were only beginning to be recognized, and paleontology and archaeology were still in their infancy. Now, there is a rich and ever-growing record from Africa, Asia, and Europe, and I have been privileged to work in one of the most exciting eras for discoveries about our origins. There have been highly significant fossil finds, of course, but there have also been remarkable scientific breakthroughs in the amount of information we can extract from those finds. In this first chapter I will outline the evidence that has been used to reconstruct where our species originated, and the very different views that have developed, including my own. There are in fact two origins for modern human features that we need to consider. Here, I will talk about our species in terms of the physical features we humans share today, for example, a slender skeleton compared to our more robust predecessors, a higher and rounder braincase, smaller brow ridges, and a prominent chin. But there are also the characteristics that distinguish different geographic populations today—the regional or "racial" characteristics, such as the more projecting nose of many Europeans, or the flatter face of most Orientals. I will discuss their quite different origins later in the book.

In *The Descent of Man* (1871), Darwin suggested that Africa was the most likely evolutionary homeland for humans because it was the continent where our closest relatives, the African apes, could be found today. However, it was to be many years before the fossil evidence that was ultimately to prove him right began to be discovered. Before then, Europe with the Neanderthals, "Heidelberg Man," and the spurious "Piltdown Man," and Asia with "Java Man," were the foci of scientific attention concerning human ancestry. But the 1921 discovery of the Broken Hill skull in what is now Zambia, and the 1924 discovery of the Taung skull (from South Africa), started the process that gave Africa its paramount importance in the story of human evolution, even if that process still had many years to run. By the 1970s a succession of fossils had established that Africa not only was the place of origin for the human line (that is, the continent in which the last common ancestor of humans and chimpanzees lived) but was probably also where the genus *Homo* (humans) had originated. But where did our own species, *Homo sapiens* (modern humans), originate? This was still unclear in the 1970s and remained so until quite recently.

When Charles Darwin wrote in the *Origin of Species*, "light would be thrown on the origin of man and his history," he was reluctant to say any more on the subject, as he admitted twelve years later in the introduction to *The Descent of Man*: "During many years I collected notes on the origin or descent of man, without any intention of publishing on the subject, but rather with the determination not to publish, as I thought that I should thus only add to the prejudices against my views." But in the intervening years he had been fortified by a growing number of influential supporters and thus felt ready—finally—to tackle the controversial topic of human origins. He then went on to say: "The sole object of this work is to consider, firstly, whether man, like every other species, is descended from some pre-existing form; secondly, the manner of his development; and thirdly, the value of the differences between the so-called races of man." However, Darwin acknowledged that there were still many doubters, something that unfortunately remains as true today as it was then: "It has often and confidently been asserted, that man's origin can never be known: but ignorance more frequently begets confidence than does knowledge: it is those who know little, and not those who know much, who so positively assert that this or that problem will never be solved by science."

Darwin then proceeded to pay tribute to a number of other scientists for their work on human origins, particularly the German biologist Ernst Haeckel, and this is especially interesting as Haeckel differed from him and Thomas Huxley ("Darwin's bulldog") over a critical question about our origins, a question that continues to be debated even today. In *The Descent of Man* Darwin wrote: "We are naturally led to enquire, where was the birthplace of man at that stage of descent when our progenitors diverged from the catarrhine stock [the catarrhines group includes apes and monkeys]? . . . In each great region of the world the living mammals are closely related to the extinct species of the same region. It is therefore probable that Africa was formerly inhabited by extinct apes closely allied to the gorilla and chimpanzee; and as these two species are now man's nearest allies, it is somewhat more probable that our early progenitors lived on the African continent than elsewhere." However, he then proceeded to caution, "But it is useless to speculate on this subject . . . as there has been ample time for migration on the largest scale."

Not only did Darwin have to deal with a dearth of fossil evidence in 1871, including a complete absence of any humanlike fossils from Africa, but there was also no knowledge of the concept of continental drift (the idea that landmasses migrated in the past, splitting and realigning as they moved across the Earth's surface). This process is now known to underlie many of the present distributions of plants and animals (for example, the unique assemblages of species found in places like Australia and New Zealand). Previously, to explain puzzling links between species in different regions, now-sunken continents were often postulated. For example, lemurs are rather primitive primates that today are found only on the island of Madagascar, some three hundred miles off the coast of Africa, but ancient lemurlike fossils had been found in the Indian subcontinent, and such similarities led the British zoologist Philip Sclater to hypothesize in 1864 that there was once a large continent, which he named Lemuria, stretching across much of what is now the Indian Ocean.

Using the concept of this lost continent, Haeckel argued for a different ancestral homeland for humans: "There are a number of circumstances which suggest that the primeval home of man was a continent now sunk below the surface of the Indian Ocean, which extended along the south of Asia . . . towards the east; towards the west as far as Madagascar

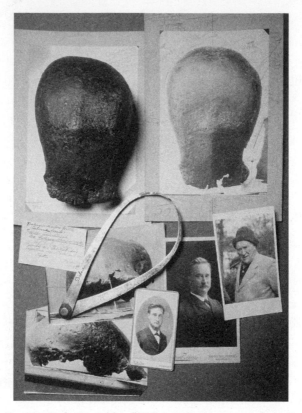

Eugène Dubois and his "*Pithecanthropus erectus*" skull.

and the southeastern shores of Africa. By assuming this Lemuria to have been man's primeval home, we greatly facilitate the explanation of the geographical distribution of the human species by migration." Moreover, Haeckel differed from Darwin and Huxley in favoring the gibbon and orangutan of southeast Asia as better ape models for human ancestry than the gorilla and chimpanzee of Africa. And whereas Darwin followed the geologist Charles Lyell in arguing that the fossil record of human evolution was still unknown because the right regions had not yet been searched (in particular Africa), Haeckel preferred the explanation that most of the critical evidence was now sunk beneath the Indian Ocean.

During Darwin's lifetime, the Neanderthals were already known from their fossil remains as ancient inhabitants of Europe. While some scientists pushed them into the position of "missing links," reconstructing them with bent knees and grasping big toes, others like Huxley recognized them as big-brained, upright, and unmistakably human. Darwin never lived to see the first discovery of a really primitive human fossil, announced by a Dutch doctor, Eugène Dubois, in 1891. Dubois had been inspired by Haeckel's writings to get an army posting to what was then the Dutch East Indies (now Indonesia), to search for ancient remains. Haeckel had created the name "*Pithecanthropus alalus*" ("Ape Man without Speech") for a hypothetical link between apes and humans that he believed had once lived in Lemuria. Dubois was blessed with luck in his excavations on the island of Java and soon found a fossilized and apelike skullcap and a human-looking thighbone. He named these "*Pithecanthropus*" (in honor of Haeckel) "*erectus*" (because the femur indicated this creature walked upright, as we do). We now know this species as *Homo erectus*, a wide-ranging and long-lived species of early human. But because this first find of the species was made on the Indonesian island of Java, it tended to reinforce Haeckel's and Dubois's notions of a Lemurian/southern Asian origin for humans, rather than an African one.

In naming "*Pithecanthropus erectus*," Dubois was following the system laid down over a century earlier by that greatest of all classifiers, the Swedish naturalist Carl Linnaeus. The Chinese sage Confucius said that it was "a wise man" who specified the names of things, and by happy coincidence this was the name, in Latin, that Linnaeus chose for the human species: *Homo sapiens*. Before Linnaeus there were many different ways of naming and grouping plants and animals, often based at random on particular features that they showed—color, say, how they moved around, or what they ate. But Linnaeus believed in grouping living things by the bodily features they shared, and at the heart of his system were the two names applied to every natural kind, or species: its group or genus name capitalized, and its particular species name. Thus *Homo* ("Man") and *sapiens* ("wise"). The system is a bit like a surname (the genus name *Homo*) and a first name (differentiating the different children with a particular surname, in our case *sapiens*). In the most-cited tenth edition of his book *Systema Naturae* (1758) he also named four geographic subspecies: "*europaeus*," "*afer*," "*asiaticus*," and "*americanus*," introducing some dubious anecdotal behavioral distinctions in line with then current European notions about the superiority of the European

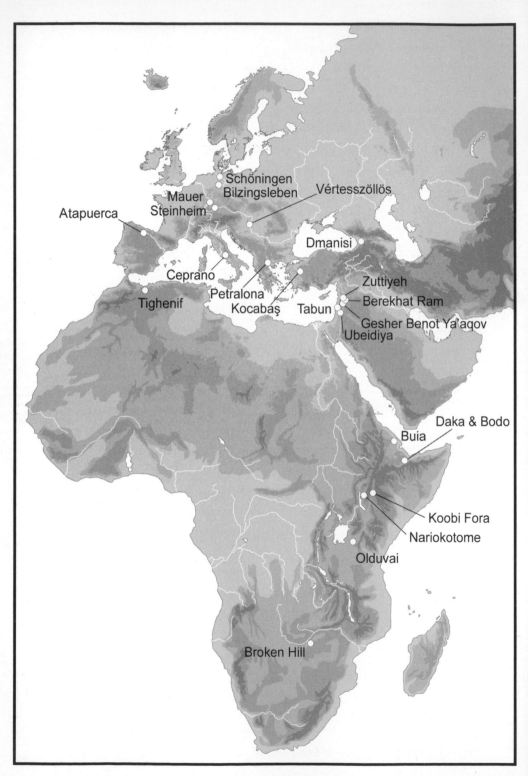

Map showing early human sites.

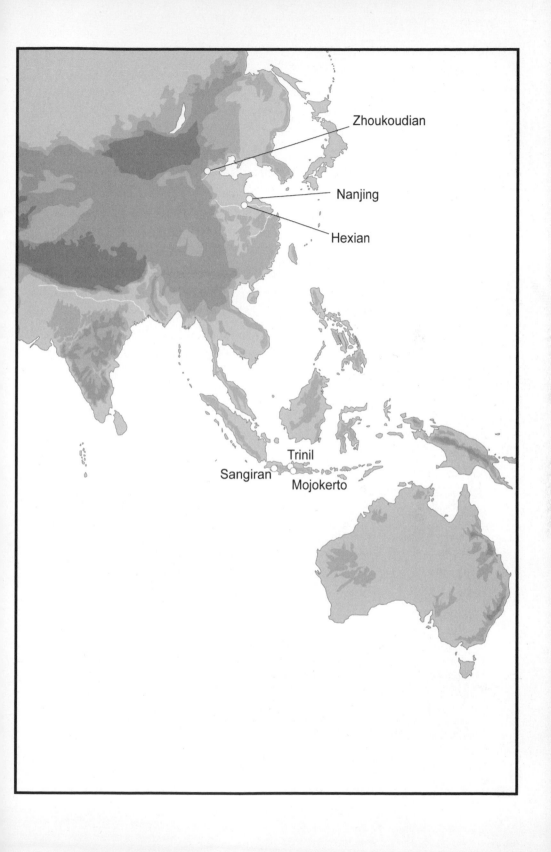

subspecies. For example, while "*europaeus*" was, of course, governed by laws, "*americanus*" was governed by customs, "*asiaticus*" by opinions, and the African subspecies "*afer*" by impulse.

In the early 1900s, evidence continued to accumulate in favor of a non-African origin for humans, and the focus returned to Europe. Further Neanderthal remains were found in Croatia and France, and a more ancient and primitive fossil jawbone was unearthed in the Mauer sandpit near Heidelberg in Germany in 1907. As enough material began to accumulate, scientists started to build evolutionary trees from the fossil evidence. These tended to fall into two main categories: ones where the fossils were arranged in a linear sequence leading from the most primitive form (for example, Java Man or Heidelberg Man) to modern humans, with few or no side branches (like a ladder); and others (like a bush) where there was a line leading to modern humans, and the other fossils with their primitive features were placed in an array of side branches leading only to extinction.

The combination of Darwin's and Wallace's publications on the transmutation of species and a proliferating Pleistocene fossil human record led to the expectation that there must have been many more ancient

A replica of the jawbone unearthed in the Mauer sandpit near Heidelberg in Germany in 1907, together with one of the Boxgrove incisor teeth.

species of humans (the Pleistocene is a recent geologic epoch, poorly dated during Darwin's time, but now believed to stretch from about 12,000 to 2.5 million years ago). William King had named the first fossil-based species *Homo neanderthalensis* in 1864, from the skeleton discovered in the Neander Valley in 1856. Within fifty years, the new European finds were being assigned to dozens of new human species in an unfortunate tumult of typology, where trivial differences were elevated to assume real biological significance. Thus, the completely modern-looking remains that had been found in the sites of Cro-Magnon, Grimaldi, Chancelade, and Oberkassel became the human species "*spelaeus*," "*grimaldii*," "*priscus*," and "*mediterraneus*," respectively, while the remains from Spy, Le Moustier, and La Chapelle-aux-Saints became "*spyensis*," "*transprimigenius*," and "*chapellensis*," despite their resemblance to the remains already designated *H. neanderthalensis* from the Neander Valley. This trend for what we can call extreme "splitting" continued up to about 1950, when the pendulum swung back to the opposing tendency to "lump" fossils together in just a few species.

Suggestions that Europe may have hosted even more primitive human relatives started to emerge from a gravel pit at Piltdown in southern England in 1912, giving rise to yet another species called "*Eoanthropus dawsoni*" ("Dawn Man of Dawson"—Charles Dawson being the principal discoverer). Parts of a thick but large-brained skull, coupled with a distinctly apelike jaw, turned up there with ancient animal fossils and primitive stone tools, suggesting an age as great as that of Java Man. Africa had nothing to compare with these burgeoning finds, but that finally began to change in the 1920s. However, circumstances were such that these first finds still failed to switch the focus of human origins to Africa.

The Broken Hill (Kabwe) skull, discovered in 1921, was the first important human fossil from Africa, but it was a puzzling find. Although it was assigned to the new species "*Homo rhodesiensis*" by Sir Arthur Smith Woodward of the British Museum, the Czech-American anthropologist Aleš Hrdlička dubbed it "a comet of man's prehistory" because of the difficulty in deciphering its age and affinities. The skull was found in cave deposits that were being quarried away during metal ore mining, in what is now Zambia (then the British colony of Northern Rhodesia). It's one of the most beautifully preserved of all human fossils, but it displays a strange mixture of primitive and advanced features, and its face is dominated by an enormous brow ridge glowering over the eye sockets. And because it was found during quarrying, which eventually destroyed the whole Broken Hill mine, its age and significance remain uncertain even today (but see the final chapter for the latest developments).

Three years later an even more primitive find was made in a limestone quarry at Taung, South Africa—a skull that looked like that of a young ape. It was studied by a newly established professor of anatomy in Johannesburg, named Raymond Dart, and in 1925 he published a paper in the scientific journal *Nature*, making some remarkable claims about the fossil. He argued that it showed a combination of ape and human features, but that its teeth, brain shape, and probable posture were humanlike. Dart named it *Australopithecus africanus* ("Southern Ape of Africa"), and he declared that it was closely related to us, and even a potential human ancestor. Dart's claims were treated with great skepticism by the scientific establishment, particularly in England. This was partly because of judgments about Dart's youth and relative inexperience, and partly because the fossil was that of a child (young apes may look more "human" than adult apes). Others thought that the finds from Java, Heidelberg, and Piltdown provided much more plausible ancestors than *Australopithecus africanus*. And finally, the location and estimated age of Taung also counted against it.

No one (not even Darwin and Huxley) had considered southern Africa as a location for early human evolution, and as the Taung skull was guessed to be only about 500,000 years old, it was thought too recent to be that of a genuine human ancestor. Instead, it was considered to represent a peculiar kind of ape, paralleling humans in some ways. We now know, of course, that the australopithecines represented a long and important phase of human evolution that lasted for over 2 million years, and which is recognized at sites stretching from Chad in the Sahara to many more in eastern and southern Africa. And we have also known since their exposure in 1953 that the misleading Piltdown remains were fraudulent and had nothing to do with our ancient ancestry.

Other finds made at this time continued to keep the focus outside of Africa, and those made in cave deposits at Zhoukoudian near Beijing from 1921 onward began to reveal a Chinese counterpart to Java Man initially dubbed "*Sinanthropus pekinensis*" ("Chinese Man of Peking"). Systematic excavations carried out from 1927 until the present day have yielded many skull and body parts of humans who lived there about half a million years ago, people who resembled the growing collection of fossils from Java closely enough for them to be eventually grouped in the single species *Homo erectus*. This species is a crucial one for studies of our origins, because it's at the heart of radically different views of our evolution that have emerged over the last seventy years or so. Most anthropologists recognize the existence of at least two human species during the last million years—the extinct *Homo erectus* and our own species, *Homo sapiens*—but there are very different views on how these species are related.

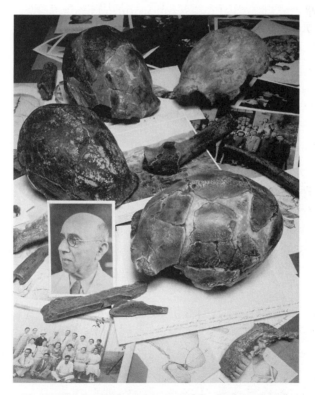

Franz Weidenreich and some of the "Peking Man"
fossils of *Homo erectus* that inspired him to create
an early version of the Multiregional model of
human origins.

What is now known as the Multiregional model of modern human origins was first proposed in the 1930s by Franz Weidenreich, a German anthropologist, who based many of his arguments on studies of the Zhoukoudian *Homo erectus* fossils. Weidenreich suggested that *Homo erectus* gave rise to *Homo sapiens* across its whole range, which, about 1 million years ago, included Africa, China, Indonesia, and perhaps Europe. In his view, as the species dispersed around the Old World (it's not known from regions such as Australia and the Americas), it developed the regional variation that lies at the roots of modern "racial" differentiation. Particular features in a given region persisted in the local descendant populations of today. For example, he argued from the fossils that Chinese *Homo erectus* specimens had the same flat faces and prominent cheekbones as modern oriental populations, while Javanese *Homo erectus* had robustly built cheekbones and faces that jutted out from the braincase, characteristics argued to be especially marked in modern Australian Aborigines.

At the other extreme from Weidenreich's Multiregional model was the view that the special features of modern humans (such as a high forehead, a chin, and a slender skeleton) would have required a long time to evolve, and hence the line leading to *Homo sapiens* (the "pre-*sapiens*" lineage) must have been very ancient and developed in parallel with large-browed and robust forms such as *Homo erectus* and the Neanderthals. This is an old idea, which came to prominence early in the twentieth century through influential researchers like Marcellin Boule (France) and Arthur Keith (United Kingdom), and aspects of it were taken up later by Louis Leakey, working in Kenya and Tanzania. The supporting evidence came and went through the last century, including at times specimens like Piltdown and the modern-looking Galley Hill skeleton from Kent— the former now known to be a fake and the latter wrongly dated.

Between the extremes of Multiregionalism (which potentially included every human fossil in our ancestry) and the Pre-*sapiens* model (which excluded most of them), there were intermediate models, ones which featured early Neanderthals in the story. The critical fossils this time were from Mount Carmel in what was then Palestine. They were discovered by an international expedition excavating a series of caves near Haifa during the late 1920s and 1930s. In two of the caves, Skhul and Tabun, they found human fossils that had apparently been intentionally buried. Moreover,

Louis Leakey with the Olduvai Gorge "*Zinjanthropus*"
skull, which his wife, Mary, discovered in 1959.
It was the first important fossil to be dated by the
potassium-argon method.

they were associated with the kinds of stone tools that in Europe were
associated with the Neanderthals. And yet the fossils seemed to show
mixtures of Neanderthal and modern characteristics, so how should they
be interpreted? In the 1930s there were no accurate methods of dating
available, and so the Tabun and Skhul fossils were assumed by their
describers, Theodore McCown and Arthur Keith, to be roughly contemporaneous with each other. Some suggested that the finds might represent
hybrids between moderns and Neanderthals, but McCown and Keith
preferred to regard them as members of a single but variable ancient population, perhaps one close to the divergence of the Neanderthal and modern
lines. (In fact Keith could not quite abandon his pre-*sapiens* leanings and
thought that they were still probably off the line leading to us, because of
their Neanderthal features.)

But others saw them as evidence for a pre-Neanderthal rather than
pre-*sapiens* ancestry for modern humans, with the late or "classic" Neanderthals subsequently heading off the main line to the sidings of extinction.

Following this line of argument, the American paleoanthropologist F. Clark Howell developed a neat scenario during the 1950s where "unspecialized Neanderthals" about 100,000 years ago became isolated in Europe by the last Ice Age and evolved away from *Homo sapiens*. At the same time, those in the Middle East (such as Tabun) evolved toward modern humans via forms like those found at Skhul. Then, to complete the story, about 35,000 years ago these Middle Eastern "proto–Cro-Magnons" migrated into Europe and replaced their European Neanderthal cousins.

In contrast to this Early Neanderthal model of modern human origins, which gave the Neanderthals at least a bit part in our evolution, there were two developments out of Weidenreich's Multiregionalism after his death in 1948 that returned the Neanderthals to a central role in our evolution, and in one case even extended their role globally. The American anthropologist Carleton Coon used new fossil material to develop a comprehensive global scheme of the evolution of five different lineages of *Homo erectus*, two in Africa, and one each in Europe, China, and Australia. These five lineages evolved largely independently to become what Coon regarded as the modern races of *Homo sapiens*: "Capoid" (the Bushman of South Africa and related peoples), "Negroid," "Caucasoid," "Mongoloid," and "Australoid."

In this respect, Coon differed fundamentally from his mentor, since Weidenreich considered human evolution to consist of a network of lineages constantly exchanging genes and ideas, whereas Coon was quite frank about the divided lineages and the implications of their inferred different rates of evolution: "Wherever *Homo* arose, and Africa is at present the most likely continent, he soon dispersed, in a very primitive form, throughout the warm regions of the Old World . . . If Africa was the cradle of mankind, it was only an indifferent kindergarten. Europe and Asia were our principal schools."

The American paleoanthropologist C. Loring Brace gave Weidenreich's ideas a distinctly Neanderthal twist by arguing that *Homo erectus* evolved to modern humans in each part of the populated world by passing through a "neanderthaloid" phase. In essence, according to Brace, the Neanderthals and equivalent ancient people across the inhabited world used their front teeth as tools for manipulating food and materials, and this is what produced their especially prominent midfaces, large incisor teeth, and distinctive skull shape. When more advanced tools of the Upper Paleolithic (Upper Old Stone Age) were invented about 35,000 years ago, demands on the teeth and jaws were lifted, and so the face and skull were transformed into the shape we have today.

These were the main ideas about the origin of modern humans that I set out to test when I began my studies for a Ph.D. at Bristol University in 1970: the global Multiregional model and its Braceian development, the Early Neanderthal model, the Pre-*sapiens* model (with no place for *erectus* or Neanderthals), and one rather vaguer scheme developed by the anthropologists Bernard Campbell and Joseph Weiner called the Spectrum Hypothesis. This argued that ancient humans had different blends of what would become modern human characteristics, and they contributed in part, and differentially, to the evolution of *Homo sapiens*. So in a sense the Spectrum Hypothesis was multiregional, but some lineages had a much greater contribution to our ancestry than others. A bit of a revolution was under way around 1970, as computing power started to increase and began to influence the biological sciences. Most analyses of human evolution up to 1970 were based on direct observation, and where measurements of a fossil were taken, these were usually compared individually or through an index of just two measurements. However, multivariate programs were becoming available that could look at large numbers of measurements and specimens simultaneously, allowing more sophisticated studies of differences in size and shape. Such analyses were at the center of my studies for a doctorate, and in July 1971 I left the United Kingdom on a trip to museums and research institutes in ten European countries. The aim was to gather as many data as possible on the Neanderthals and their modern-looking successors in Europe, the Cro-Magnons, to see whether the evolutionary pattern was one of continuity or rupture. I only had a modest grant from the Medical Research Council for a four-month trip, and so I drove my old car, sleeping in it, camping, or staying in youth hostels—in Belgium I even spent one night in a shelter for the homeless. I survived many adventures, including several border confrontations and two robberies, but by the end of my 5,000-mile trip I had collected one of the largest data sets of Neanderthal and early modern skull measurements assembled by anyone up to that time.

Chris Stringer on his 1971 research trip around
Europe. It's washing day at a campsite in Yugoslavia.

Over the next two years I analyzed this information, adding comparative data on non-European fossils and modern human populations (the
latter generously supplied by the American anthropologist William Howells). The measurements were transferred to data cards and fed into a computer the size of several rooms, but which had less processing power than
my last mobile phone! Nevertheless, the results were instructive. Neanderthal skulls were no more similar to those of recent Europeans than they
were to Africans, Eskimo, or Native Tasmanians, and Cro-Magnon skulls
did not neatly slot between the Neanderthals and recent Europeans. Early
modern skulls from around the world seemed to cluster with their modern counterparts rather than with any archaic skulls from the same
regions. The former results provided no support for a Neanderthal ancestry for the Cro-Magnons, and the latter results contradicted Multiregional and Spectrum expectations. Studying the sequence in Europe
before Neanderthal times also gave no support to the Pre-*sapiens* model
either, because very early European fossils could not be divided into
modern-like and Neanderthal-like; they seemed to show the gradual
development of only Neanderthal features through time.

Things were not quite as clear in the Middle East, although there did not seem to be any "intermediate" fossils between Neanderthals and moderns there either. Skulls from Tabun and the Israeli cave of Amud seemed to be basically Neanderthal, while those from Skhul Cave seemed much more modern. But because none of these finds were well dated in the 1970s, I couldn't exclude the possibility that, given enough time, the Israeli Neanderthals could have been transformed into early moderns, in line with the Early Neanderthal model of scientists like Clark Howell. However, a surprising alternative ancestor for the Skhul and Cro-Magnon early moderns did emerge from my results. A skull discovered in 1967 in the Omo Kibish region of Ethiopia, by a team led by Richard Leakey (the son of the famous prehistorians Louis and Mary Leakey), looked very modern in my skull shape analyses, confirming the first studies by the anatomist Michael Day; yet preliminary dating work suggested it could have been as much as 130,000 years old, more ancient than most Neanderthals. And there was an enigmatic North African skull, found in the Moroccan site of Jebel Irhoud in 1961. In skull shape it seemed Neanderthal in some ways, yet its facial shape was non-Neanderthal, partly primitive and partly modern. With an age thought to be only around 40,000 years, it was difficult to fit Jebel Irhoud into any scenario, but it and the Omo skull provided clues that Africa was going to have its own story to tell, when more data came in.

As my work developed through the 1970s and early 1980s, I gravitated increasingly toward what Bill Howells in 1976 had dubbed the Garden of Eden (or Noah's Ark) model. This was named not because Howells was any kind of biblical creationist, but because of the implication that all modern human variation had developed from a single center of origin. A lack of fossils from many parts of the world, together with inadequate dating for many of those we did have, meant that neither Howells nor I could specify where that center of origin might have been, although we thought we could exclude the European and Middle Eastern territories of the Neanderthals. We both believed that the distinctive shared features of modern humans, such as the high rounded skull, small brows, and chin, implied a recent common origin, as otherwise there would have been much greater differentiation over time. And I started to move away from the then widespread idea that fossils as different-looking as Broken Hill, the Neanderthals, and Cro-Magnon should all be classified with us as variants of our species, *Homo sapiens*. Initially I agreed with some other workers in differentiating "anatomically modern *sapiens*" (such as Skhul and Cro-Magnon) from "archaic *sapiens*" forms such as Neanderthals and Broken Hill. But during the 1980s I increasingly favored limiting the *sapiens* term to fossils closely resembling us. Moreover, along with a few other heretics, I started to argue that the Neanderthals should be returned to the status granted them by William King in 1864 as a distinct species, *Homo neanderthalensis*. I also suggested that the Broken Hill skull found in 1921 could be grouped with more primitive European forms (for example, the Heidelberg jaw discovery of 1907) as *Homo heidelbergensis*.

As my views on our origins were developing toward a single-origin model, evidence began to accumulate that Africa was especially important in this story. The Omo Kibish find was joined by material from the sites of Border Cave and Klasies River Mouth Caves in South Africa. Moreover, new dating work hinted that Africa was not the backwater in cultural evolution that most considered it to be. Archaeologists such as Desmond Clark and Peter Beaumont argued that it might instead have been leading the way in the sophistication of its stone tools. By 1980 I was privately convinced that Africa was the main center of our evolution but, because of dating uncertainties, I could not rule out the Far East as also playing a role. It took another four years for me to take a strong "Out of Africa" stance publicly, as various lines of evidence started to fit together in my mind.

However, further confusion was sown by the strong reemergence of Weidenreich's Multiregional views in 1984. These were given a new lease on life by Milford Wolpoff (United States), Alan Thorne (Australia), and Wu Xinzhi (China). They distanced themselves from Coon's views by returning to Weidenreich's emphasis on the importance of gene flow between the geographic lines, considering the continuity in time and space between the various forms of *Homo erectus* and their regional descendants to be so complete that all of them should be classified with modern people as representing only one species: *Homo sapiens*. Thus in this model there was no real "origin" for the modern form of *Homo sapiens*. A feature like the chin might have evolved in a region such as Africa, and spread from there by interbreeding across the human range, followed by selection for it if it was an advantageous characteristic. Another feature such as our high forehead might have developed in, say, China, and then similarly spread from there through interbreeding. Thus modern humans could have inherited their "local" features through continuity with their ancient predecessors, while global characteristics were acquired via a network of interbreeding.

But new developments in genetics research were about to have a huge impact. In 1982 I became aware of research work on a peculiar type of DNA that is found outside the nucleus of cells, in the *mitochondria*. These are little bodies that provide the energy for each cell, bodies that probably originated from a once-separate bacterium, which somehow survived being engulfed by a primitive cell. They then coevolved to confer mutual advantage and developed into the mitochondria that most organisms have throughout their cells. In humans, the DNA of a mother's mitochondria is cloned in her egg when it becomes the first cell of her child, and little or no mitochondrial DNA from the father's sperm seems to be incorporated at fertilization. This means that mitochondrial DNA (mtDNA) essentially tracks evolution through females only (mothers to

Milford Wolpoff, an architect of Multiregionalism, with a
Homo erectus skull from Java.

daughters), since a son's mtDNA will not be passed on to his children. This type of DNA mutates at a much faster rate than normal (nuclear) DNA, as we will discuss in chapter 7, allowing the study of short-term evolution. Early work on human mitochondria seemed promising, showing that our species apparently had low diversity and a recent origin, but the geographic patterns seemed unclear as to where that origin might be. By 1986 I had heard through the grapevine that startling new mtDNA results were on the way to publication, and a year later they appeared in the science journal *Nature*, shaking up arguments about recent human evolution in such a way that things would never be the same again. This seminal publication by Rebecca Cann, Mark Stoneking, and Allan Wilson put modern human origins on the front pages of newspapers, journals, and magazines for the first time.

About 150 types of mtDNA from around the world were investigated, and their variation was determined. Then a computer program was used to connect all the present-day types in an evolutionary tree, with the most economical pattern of evolutionary change (mutations), reconstructing hypothetical ancestors for the living types. In turn, the program connected those ancestors to each other, until a single hypothetical ancestor for all the modern types was created. The distribution of the ancestors implied that the single common ancestor must have lived in Africa, and the number of mutations that had accumulated from the time of the common ancestor suggested that this evolutionary process had taken about 200,000 years. This, then, was the birth of the now-famous Mitochondrial Eve, or "lucky mother," since the common mitochondrial ancestor must necessarily have been a female. These results seemed to provide strong evidence for a Recent African Origin view for modern humans, since the research suggested that a relatively recent expansion from Africa had occurred, replacing any ancient populations living elsewhere, along with their mtDNA lineages. However, the work was soon heavily criticized. It was shown that the kind of computer program used could actually produce many thousands of trees which were all more or less as economical as the published one, and not all of these alternative trees were rooted in Africa. Moreover, other researchers criticized the calibration of the time when Mitochondrial Eve lived, while yet others questioned the constitution of the modern samples analyzed (for example, many of the "African"

samples were actually from African Americans). As a result, multiregionalists were, for a while at least, able to reject these mtDNA results as irrelevant or misleading, arguing that fossil evidence (and their interpretation
of it) remained the only valid approach to reconstructing recent human
evolution.

However, the results strongly supported the Recent African Origin
view that people like Günter Bräuer (from Hamburg) and I had been
developing from the fossils. Günter was less inclined to view *Homo sapiens*
as a newly evolved species, and more inclined to think that hybridization
had occurred with people like the Neanderthals, following the dispersal
from Africa, but we both welcomed the new mtDNA data. For me, it gave
greater confidence that even where the fossil evidence was patchier or more
ambiguous, such as the Far East and Australasia, the story of replacement
that I had read from the European record probably applied there too.

In 1987 the archaeologist Paul Mellars and I co-organized an international conference in Cambridge where recent fossil and archaeological
results were compared with the new DNA data, and the discussions were
electric at times, as experts got to grips with the rapidly changing landscape of recent human evolution. A year later, taking the conference
discussions and DNA analyses fully on board, I wrote a review of that
emerging picture for the journal *Science*, with my Natural History Museum
colleague Peter Andrews. We laid out the two contrasting models of Multiregionalism versus Recent African Origin and what would be expected
from the fossil, archaeological, and genetic data if either model was an accurate representation of recent human evolution. (I actually prefer to use the
term Recent African Origin [RAO], despite the popularity of Out of Africa,
because we know from more ancient fossils that there were earlier human
dispersals from Africa. Hence some people distinguish them as Out of
Africa 1, Out of Africa 2, et cetera, although we don't actually know how
many there were—and no doubt there were some "into Africa" events as
well!)

Overall, we showed that RAO was best supported, although we recognized that the archaeological record in general and the fossil records
of several regions in particular were still not adequate to test the models
properly. I was shocked, though, by some of the vitriolic reactions to that
paper. Both in the anonymous reviews that some other scientists sent to

Two of the architects of the Recent African Origin model, Günter Bräuer (left)
and Chris Stringer, pictured in the 1980s.

the journal before publication, and in letters and media comments afterward, scorn was poured on our views and interpretations, a scorn that
seemed to extend to personal abuse at times. Relations became strained
with a number of scientists, some of whom were people I certainly counted
as my friends. Cordiality was eventually restored in most cases, but for a
few people, what was seen as an extreme position, in league with the her-
esy of Mitochondrial Eve, was not easily forgiven or forgotten.

As more fossil and, particularly, genetic data emerged to support
a recent African origin, what we can term the classic RAO model was
developed by a number of researchers, including me, working separately
or in collaboration. By the turn of the millennium, this had become the
dominant view. Fleshing it out with the consensus view for earlier human
evolution, the classic RAO model argued for an African origin of two
human species—*Homo erectus* and *Homo sapiens*—and perhaps also of
Homo heidelbergensis between them (in my view, though, the derivation of
heidelbergensis is still unclear). Having evolved from something like the
earlier species *Homo habilis* in Africa nearly 2 million years ago (Ma),
Homo erectus dispersed from Africa about 1.7 Ma, in the event commonly
known as Out of Africa 1. The species spread to the tropical and subtropical regions of eastern and southeastern Asia, where it may have lingered
on, evolved into other forms, or died out. About 1.5 Ma, African *erectus*
developed more advanced stone tools called *handaxes*, but these did not

spread far from Africa until they turned up rather suddenly with the descendant species *Homo heidelbergensis* in places like southern Europe, and then in Britain, 500,000 to 600,000 years ago.

So my view was that *H. heidelbergensis* subsequently underwent an evolutionary split around 300,000 to 400,000 years ago: it began to develop into the Neanderthals in western Eurasia, while the line in Africa had evolved into the ancestors of modern humans by about 130,000 years ago. The origin of modern *Homo sapiens* must have been a relatively recent and restricted one in Africa, based on marked similarities between recent humans in both body form and DNA, and it may have been quite rapid, in one small favored area such as East Africa. Some modern humans dispersed to the Middle East (Israel) about 100,000 years ago, and they had perhaps moved on as far as Australia by about 60,000 years. However, *Homo sapiens* did not enter Europe until about 35,000 years ago, following the rapid development of more advanced Later Stone Age tools and complex behaviors by African moderns about 50,000 years ago. Such progress finally allowed the moderns to spread into Europe, where, as Cro-Magnons making Upper Paleolithic tools, they quickly took on and replaced the Neanderthals through their superior technology and adaptations. Bear this narrative in mind, as I will revisit it at various times later in this book.

If RAO is the most accurate model, regional ("racial") variation only developed during and after the dispersal from Africa, so any seeming continuity of regional features between *Homo erectus* and present counterparts in the same regions outside of Africa must have been as a result of parallel evolution or coincidence, rather than of genes passed down from archaic predecessors, as argued in the Multiregional model. Like that model, RAO argued that *Homo erectus* evolved into new forms of humans in inhabited regions outside of Africa, but in RAO these non-African lineages eventually became extinct, without evolving into modern humans. Some, such as the Neanderthals, must have been replaced by the spread of modern humans into their regions, and hence the RAO model not only is popularly known as Out of Africa but is sometimes also known as the Replacement model.

As RAO gathered support and influence, it increasingly made an impact on the views of people like the American anthropologists Fred Smith and Erik Trinkaus, who believed in continuity outside of Africa but were not classic multiregionalists. Instead, they advanced what has become known as the Assimilation model, which can be seen as a moderate position between the extremes of RAO and what I have dubbed classic Multi-regionalism: one where Africa dominated as the source of modern features, but where these were taken up more gradually by people outside of the continent, through a blending of populations. Modern features thus diffused out of Africa rather than being imposed through the invasion and dominance of dispersing moderns, and early moderns outside of Africa could therefore be expected to show features of the "natives" with whom they were mixing. And while the various models of human evolution were adjusting themselves to the post-mtDNA landscape, the genetic work itself was undergoing reevaluations.

I already mentioned the heavy criticisms of the 1987 "Eve" paper, from the point of view of the samples used, the methods of analysis, the rate of evolution, and the strong conclusions drawn. The team involved in the original work acknowledged that there were deficiencies, and, over the next few years, they set out to address the problems in a series of further analyses that served only to reinforce their conclusions, as we shall see in chapter 7. But as we shall also observe, most workers now agree that mtDNA, while very useful, is only one small part of the genetic evidence we need to reconstruct our evolutionary origins.

For the rest of this book, I will mainly be discussing three other human species along with our own: *Homo erectus*, *H. heidelbergensis*, and *H. neanderthalensis*. So how do we recognize distinct human species in the fossil record and our own ancestors? Well, that is not a straightforward question, and specialists will give differing answers. (For example, as I explained earlier, multiregionalists often regard *Homo sapiens* as the only human species on Earth during the last million years, so species like *Homo erectus* and *Homo heidelbergensis* have no real meaning for them.) But for me, there are features in the skeleton that, taken together, can diagnose distinct human species in the past, and that similarly characterize our species today. Because of variations in time and space, these features are rarely absolute, but in combination I think they can distinguish separate evolutionary lines that we can call species, based on their skeletal structure.

For our own species, *Homo sapiens* (modern humans), these features include: a large brain volume; neurocranial globularity (the curvature and doming of the bones of the braincase, and its increased height); in rear view a braincase that is wider at the top and narrower at the base; a higher and more evenly arched temporal bone at the side of the braincase; decreased height of the face and its tucking-in under the braincase; a small and divided brow ridge; a narrow area of bone between the eye sockets; increased projection of the middle of the face and nose; a bony chin on the lower jaw, present even in infants; simplification and shrinkage of tooth crowns; a lightly built tympanic bone (this contains the ear bones); a short pubic ramus that is nearly circular in cross section (this is a bone at the front of our pelvis); no iliac pillar (this is a near-vertical ridge of bone reinforcing the pelvis, above the hip socket); and femora (thighbones) that are oval in cross section and thickened most at the front and back.

In contrast, for *Homo erectus*, the human species that had appeared in Africa and Asia more than 1.5 million years ago, the characteristics included a small average brain volume; a relatively long and low braincase, narrow across the top but broad across the base; a lower and more triangular temporal bone; an angled occipital bone at the back of the skull, with a strong torus (ridge of bone) across it; bony ridges that reinforce the frontal and parietal bones of the braincase; a thick tympanic bone; a strong and continuous supraorbital torus (brow ridge); a strong postorbital constriction (the skull is pinched in behind the brow ridge when viewed from above); a wide area of bone between the eye sockets; a face that juts out from the braincase; a flatter and elongated superior pubic ramus; an iliac pillar; and femora that are rounded and evenly thickened in cross section.

Homo erectus seems primitive in many respects by the standards of later humans, but it represented a benchmark of change to the human condition in many aspects of its skeleton: a brain size beyond any ape or australopithecine, a human face with projecting nasal bones, small teeth, a humanlike posture for the skull, and a body frame of human rather than apelike proportions. The evolutionary biologists Dennis Bramble and Daniel Lieberman believe that *erectus* had made a fundamental transition to life in the open, first scavenging and then hunting over long distances. We are unique among primates in our capacity for endurance running, which may first have evolved to allow humans to get to carcasses for scavenging ahead of the competition. And people like the San today are able to gradually wear down their prey through persistent pursuit: ungulates, for example, can run much faster than humans over short distances but completely exhaust themselves over long distances, at which point they are easy to dispatch. Features of the *erectus* (and later human) skeleton in body shape, legs, ankles and feet, head balance and stability, and our reliance on sweating to thermoregulate could all be relics of an early adaptation to sustained running, according to Bramble and Lieberman.

H. heidelbergensis, present in Africa and Europe more than 500,000 years ago, shows combinations of features found in the more primitive *erectus* fossils and those found in the later Neanderthal and modern *sapiens* fossils, as befits a possible intermediate species: a brow ridge like *erectus*, but often filled with extensive sinuses (voids); an occipital bone like that of *erectus*; a wide interorbital breadth like *erectus*; a superior pubic ramus like *erectus*; an iliac pillar like *erectus*; rounder femora like *erectus*; brain volumes that overlap the smaller values of *erectus* and the larger ones of *H. sapiens* and *H. neanderthalensis*; a braincase higher than *erectus*, and parallel-sided in rear view; a face intermediate between *erectus* and later humans in its overall projection from the braincase; a temporal bone more like those of *sapiens* and *neanderthalensis*; a tympanic like Neanderthals and moderns; increased projection of the middle of the face and nose (as in Neanderthals and moderns); and, in some cases, inflated cheekbones that retreat at the sides, like those of Neanderthals.

The Neanderthals are advanced humans and thus share features with both *heidelbergensis* and with us. Yet there are also some retained primitive traits and those that betoken a separate evolutionary pathway. They have an elongated superior pubic ramus like *erectus* and *heidelbergensis*; rounder femora like *erectus* and *heidelbergensis*; a large brain volume like ours; a high and arched temporal bone like ours; reduced interorbital breadth; reduced total facial projection; a lightly built tympanic; in many Neanderthals, simplification and shrinkage of tooth crowns as in *sapiens*; weak or absent iliac pillar.

Then there are the features that seem to distinguish the Neanderthals

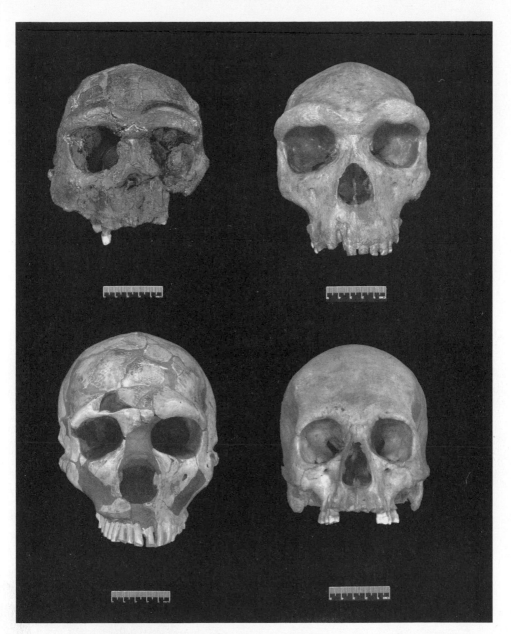

(*Clockwise from top left*) Skulls of *erectus* (Sangiran, Java), *heidelbergensis* (Broken Hill, Zambia), *sapiens* (Indonesia), and *neanderthalensis* (La Ferrassie, France).

as an evolutionary lineage. Some of these are concerned with a distinctive body shape, rib cage, and limb proportions, but the clearest ones are on the skull: a double-arched brow ridge with central sinuses; a double-arched but small occipital torus with a central pit (the suprainiac fossa); a spherical vault shape in rear view; distinctive shape of the semicircular canals of the inner ear (see chapter 3); strong midfacial projection and cheekbones that are inflated and retreat at the sides; a high, wide, and projecting nose; large and nearly circular orbits; a high but relatively narrow face; and enlarged front teeth (incisors), which are hollowed (shoveled) on the inside surfaces of the upper centrals.

The feature that stands out (literally) in these comparisons of modern

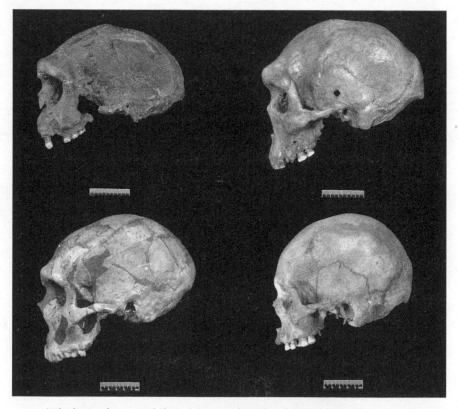

(*Clockwise from top left*) Side view of skulls of *erectus* (Sangiran, Java),
heidelbergensis (Broken Hill, Zambia), *sapiens* (Indonesia),
and *neanderthalensis* (La Ferrassie, France).

and archaic species is the strong brow ridge of the latter, and its absence in the former. The anatomist Hermann Schaaffhausen, one of the first describers of the original Neanderthal skull, called its strong brows "a most remarkable peculiarity," and although there have been many scientific hypotheses to explain their presence or absence, none really convince me. The fact that many of the huge brow ridges in fossils are hollowed inside, with large sinuses (air spaces), suggests that they are not there to bear or transmit physical forces from blows to the head or heavy chewing. The eccentric anthropologist Grover Krantz even strapped on a replica brow ridge from a *Homo erectus* skull for six months to investigate its possible benefits, finding that it shaded his eyes from the sun, kept his long hair from his eyes when he was running, and also scared people out of their wits on dark nights. For me, that last clue might be significant, and like the paleontologist Björn Kurtén I think it may even have had a signaling effect in ancient humans, accentuating aggressive stares, especially in men. Thus its large size could have been sexually selected through the generations, a bit like antlers in deer. But if that is so, why don't we have large brows like our predecessors? Well, I think the rest of this book will show that modern humans have developed so many other ways to impress each other, from weapons to bling, that perhaps the selective benefits of large brows wore off in the last 200,000 years.

If there were, in fact, different human species in the past, could they have interbred? In my view, RAO has never precluded interbreeding between modern and archaic people during the dispersal phase of modern humans from Africa. This is undoubtedly one of the main areas of confusion in studies of modern human origins: how to recognize species in the fossil record, and what this signifies. Some researchers argue that many distinct morphological groups in the fossil record warrant specific recognition, with the existence of at least ten such species of the genus *Homo* during the last 2 million years (that is, *Homo ergaster, erectus, georgicus, antecessor, heidelbergensis, rhodesiensis, helmei, floresiensis, neanderthalensis, sapiens*).

At the other extreme, some multiregionalists argue that only one species warrants recognition over that period: *Homo sapiens*. An additional complication is that different species concepts may become confused;

for example, some multiregionalists have applied what is called the *Biological Species Concept* (*BSC*) to the fossil record to justify their belief that *H. neanderthalensis* and *H. sapiens* must have belonged to the same species and would have been fully interfertile. This concept, developed from the study of living organisms, argues that a species consists of the largest community of a group of plants or animals that breeds among itself, but not with any other community. Thus it is "reproductively isolated" with reference to other species, but its own varieties can interbreed with each other. Living *Homo sapiens* would be a good example of this, since people from all over the world are potentially able to mate and have fertile children, but we are apparently reproductively quite distinct from our ape cousins. I say "apparently" because there are persistent rumors that in the 1940s and 1950s scientists in the United States and/or the Soviet Union conducted unethical experiments impregnating female chimpanzees with human sperm—the results of which, so the rumors go, have been suppressed.

And what if we could still meet a Neanderthal—could modern humans interbreed with one? First this brings up the potential conflict between the BSC (which relates to living species) and the completely different concepts that I just used to recognize species in the fossil record, such as the degree of variation in the skeleton. Using the latter measure (a morphological species concept based on what is preserved in the fossil record), I and many other anthropologists recognize the Neanderthals as specifically distinct from *Homo sapiens*. But there is a conflict at the heart of the BSC: the fact that many closely related mammal species can hybridize and may even produce fertile offspring. Examples are wolves and coyotes, bison and cows, chimps and bonobos, and many species of monkey. So we have to recognize that species concepts are humanly produced categories which may or may not always work when compared with the reality of nature. So in my view, even if there was Neanderthal-modern hybridization (and I will discuss that thorny question in chapter 7), it would not necessarily mean that Neanderthals belonged to the same species as us—it would depend on the scale and impact of the interbreeding.

Fossils—the relics of ancient species—sparked my interest in the distant past when I began collecting them as a boy, and they still fascinate me. But on their own they are just mineralized and inert bones and teeth. In the next two chapters I will show how a range of exciting new techniques are helping us return these inanimate fossils to their ancient environments and bring them back to life.

2

Unlocking the Past

JUST DOWN THE CORRIDOR FROM MY ROOM AT THE MUSEUM, locked in their own special cabinet, are some of the most notorious relics in the story of human evolution, mentioned already in chapter 1—Piltdown Man. They were found and announced, to an unsuspecting world, about a hundred years ago, and they provide a sobering reminder to all scientists to beware of something that seems too good to be true—because it may well not be true! British paleoanthropologists of the time had seen German, Dutch, and French scientists discover fossils of possible ancient ancestors, but Britain had nothing to compare with these. Moreover some of these British experts were, as we have seen, supporters of the view that our species had a deep and separate evolution from people like Java Man and Neanderthal Man. Imagine their delight, then, when a "missing link" was apparently discovered in their own backyard, in the county of Sussex. It seemed to have an apelike jawbone and a very human braincase, and these were combined to make the apeman called "*Eoanthropus dawsoni.*" Of course we now know that its "ape" jaw and "human" skull were exactly that—two completely different and relatively recent specimens maliciously combined to create a misleading transitional fossil. But the hoaxer or hoaxers were knowledgeable enough to not just rely on anatomy to fool the experts—they knew enough about how fossils were dated in 1912 to also misuse that knowledge, to make a case that the Piltdown assemblage of bones and stone tools were as ancient as the remains of Java Man. They were able to get away with it because none of the physical dating techniques that I will discuss in this chapter (such as radiocarbon dating) were known a hundred years ago, and, instead, human fossils could really only be *relatively* dated—that is, dated in relation to the material found alongside them. The hoaxer(s) planted genuine fossils of primitive mammals from other sites alongside the remains of Piltdown Man, so they would look suitably ancient. In 1953 the whole sorry story began to unfold, and when radiocarbon dating was finally applied to show the ape and human remains were both less than a thousand years old, it was the ultimate nail in Piltdown Man's coffin!

So, in this chapter, I will show how new dating techniques have revolutionized our view of human evolution in each of the main regions and time periods in which they have been applied, and I will use a variety of examples to look at the way the records of past climates and environments are influencing the story of our evolutionary origins. We now think that Neanderthals and modern humans evolved along parallel paths, the former lineage north of the Mediterranean and the latter south of it, in Africa. After several false starts, modern humans finally emerged from Africa and spread along Asian coastlines toward China and Australia. But Europe, perhaps the last bastion of the Neanderthals, seems to have remained beyond modern reach until about 45,000 years ago. We have only recently dated some of the most important human fossils, work that has revolutionized the time scale of our evolution. Fascinating new environmental and archaeological evidence also shows the complexity of the process of our evolution, and of the extinction of our close relatives the Neanderthals.

There are two main categories of dating: *relative* and *physical* (that is, based on the laws of physics, sometimes also called *radiometric* or *absolute*) *dating*. The first relates an object or layer to another object or layer in time; one may be younger than the other, or (within the limits of the method) they may be about the same age. The geologic law of superposition supposes that, unless there has been major disturbance, a layer in a geologic sequence is always younger than the layer below it; this is the main principle at work in relative dating. More rarely, a geologic event such as a tsunami or a volcanic eruption can be traced across a region, and fossils or artifacts associated with that event can be assumed to be contemporary with it, and thus with each other. But such relative dating cannot tell us how old the materials in question actually are; it can only place them in relation to each other—that is, show they are relatively older, younger, or correlated (similar) in age. Thus if I dug in my garden and found Roman pottery that looked similar to pottery found at, say, Fishbourne Roman Palace in Sussex, I could assume that my finds were about the same age as those at Fishbourne; but without independent evidence of how old Fishbourne Palace was, or the pottery was, that would be as far as I could go. I could get more detailed relative dating by, say, researching the age of Roman coins found at Fishbourne, or I could attempt a physical determination by asking a specialist in luminescence dating (see discussion later in this chapter) to use physical signals within the clay of my pottery to tell me how long ago it had been fired.

So to go farther than a relative date, we need physical clocks that will tell us how far back some rocks were laid down, how long it is since an animal or plant died, or when an event happened, such as the heating of clay or flint. Many of these clocks measure time using the natural radioactive decay of isotopes. *Isotopes* are distinct atoms of substances, such as argon or carbon, that have different atomic weights (because they contain different numbers of particles called neutrons). An example of such a technique is *potassium-argon dating*, which can be used on volcanic rocks. Potassium partly consists of an unstable isotope called potassium-40, and this isotope gradually changes over many millions of years into the gas argon. When there is a volcanic eruption, the liquid lava or hot ash contains a small proportion of potassium-40, and when the lava or ash cools and solidifies, this unstable isotope of potassium begins to change into argon, such that half of it decays into argon about every 1.25 billion years (this is its half-life). Provided the volcanic eruption was sufficiently energetic to drive out any previous argon gas (usually a reasonable assumption), and provided any newly formed argon gas remained trapped in the volcanic layer once it set hard, the amount produced can be used as a natural measure of time since the volcanic rock was deposited. In one of the first and most famous applications of this technique to archaeology, lava at the base of the site of Olduvai Gorge in Tanzania was shown to be about 1.8 million years old. This caused a sensation in 1960 because it indicated for the first time just how ancient the artifacts and humanlike fossils in Bed I at Olduvai might really be, doubling their expected age at a stroke. A more recent development from potassium-argon dating is to use the decay of argon-40 to argon-39 instead, since this can be used to date single crystals of volcanic rock with much greater accuracy over the time span of human evolution.

The most famous physical dating method is *radiocarbon dating*, based on an unstable form of carbon. The method relies on the fact that radiocarbon (an isotope of carbon called carbon-14) is constantly produced in the upper layers of the Earth's atmosphere by cosmic radiation acting on the element nitrogen. This unstable form of carbon gets taken up into the bodies of living things, along with the much more common, and stable, carbon-12. However, when the plant or animal dies, no more carbon-14 is taken in, and the amount left begins to break down by radioactive decay, such that the amount present halves about every 5,700 years—a very much shorter time span than that of potassium-argon dating. So measuring the amount of carbon-14 left in, say, a piece of charcoal or a fossil bone allows us to estimate how long it is since the plant or animal concerned was alive.

In 1949 the American chemist Willard Libby and his colleagues first applied it to a sample of acacia wood from the tomb of the pharaoh Zoser (who lived nearly 5,000 years ago). Libby reasoned that since the half-life of radiocarbon was close to 5,000 years, they should obtain a carbon-14 concentration of about 50 percent of that found in living wood, and this was confirmed. That work, and much that followed, earned Libby a Nobel Prize in 1960. The method cannot be used on very ancient materials because the amount of carbon-14 left behind is too small to measure accurately, and hence radiocarbon dating becomes increasingly unreliable beyond about 30,000 years ago. Moreover, the assumption of constant production and uptake of carbon-14 is now known to be only an approximation, due to past fluctuations in cosmic rays and changes in the Earth's atmospheric circulation—thus scientists talk of dates in *radiocarbon years* rather than real (calendar) years.

This means that other methods are needed to cross-check (calibrate) the accuracy of radiocarbon dates. Several methods have been particularly useful for dates in the last 10,000 years or so, and all of them require the counting and dating of annual layers. The first uses tree rings (dendrochronology) and builds up overlaps in patterns of growth rings from timbers preserved in buildings, boats, or natural deposits, in order to establish a long sequence where the age assessed from the wood is compared with a radiocarbon date obtained on rings within the wood. A comparable method uses varves (annually deposited layers in the bottom of deep lakes), where spans of time can be measured through counting varves, and also by radiocarbon dating of plant or animal residues within the varves. Yet another method uses radiocarbon dates obtained within annual layers of ice, and this can be taken even farther since trapped bubbles of gas in the ice preserve a snapshot of the composition of the atmosphere when a particular layer was deposited. Beyond these methods, very ancient trees preserved in bogs in New Zealand hold the promise of accurately calibrating radiocarbon to beyond 40,000 years, while ancient coral terraces can be dated both by radiocarbon and by uranium-series dating (discussed later in this chapter), giving a cross-check between independent physical dating methods, each with different assumptions.

Comparisons so far suggest that radiocarbon dating, while not exact over the last 40,000 years, is quite reliable, although sometimes off by as much as 10 percent. Unfortunately, one of its least accurate phases covers the demise of the Neanderthals and much of the spread of modern humans around the world—hence the need to further refine radiocarbon dating or supplement it with other methods wherever possible, as I will explain later in this chapter.

Many technical improvements have been made in radiocarbon dating procedures since Libby's initial work. For example, he analyzed solid carbon, while later techniques convert the carbon to gas or dissolve it in solvents. The early methods also required large sample sizes to detect radiocarbon decay, so that important artifacts or bones had to have large chunks sawn off them to attempt a date—permission for which was understandably often refused by concerned museum curators. Luckily, from about 1977, the method of *accelerator mass spectrometry* (*AMS*) has increasingly taken over, and this counts individual atoms of carbon-14 directly, rather than measuring their radioactivity. So now only milligram-sized samples are needed, allowing the dating of relics as precious as the

Shroud of Turin, the Dead Sea Scrolls, the Alpine iceman "Ötzi," and the Ice Age art of the Lascaux and Chauvet caves.

A good example of the enhanced power of radiocarbon dating came when four colleagues and I investigated one of the enduring mysteries of the Paleolithic record of Britain. Representations of Ice Age art are extremely rare in Britain, and two of the only examples known (or claimed) are from Robin Hood Cave in Derbyshire, found in the 1870s, and from the town of Sherborne in Dorset. Both showed a rather similar profile of a horse engraved on a flat fragment of bone. While the Derbyshire example was discovered by prehistorians in a cave alongside Paleolithic artifacts of appropriate age (about 14,000 years old), the "Sherborne bone" was discovered in 1912 by schoolboys from the local public school, in the vicinity of a quarry from which no comparable material had ever been reported. Serious doubts were soon raised about the authenticity of the Sherborne discovery, but direct radiocarbon dating could not have been contemplated when application of the method would probably have destroyed most or all of the object. The advent of AMS dating at Oxford University allowed us, in 1995, to drill a tiny sample from it and date the bone to about six hundred years old, while microscopic studies of the engraving showed that it was probably carried out quite recently with a metal implement, rather than a flint tool. This result was in line with suggestions from one of the staff at Sherborne fourteen years after the "discovery" that a boy had probably copied the engraving from an illustration of the Robin Hood specimen in their school library, in order to play a joke on their science teacher!

But even AMS dating is not perfect, since it finds and produces a date from whatever radiocarbon is in the sample; even a small amount of contaminant radiocarbon can greatly affect an age estimate, especially when the sample is 30,000 or 40,000 years old, and only a tiny fraction of its original radiocarbon is still there. Fortunately, new preparation procedures such as *acid-base-wet oxidation* (*ABOX*) *dating* for charcoal samples and *ultrafiltration* for bone are largely overcoming the problems of contamination in dating Paleolithic materials and are giving increasingly trustworthy determinations. The advantages provided by ultrafiltration were very well demonstrated through the redating of bone samples from

Gough's Cave in Cheddar Gorge, Somerset. This is one of Britain's most spectacular tourist caves but also one of our most important Upper Paleolithic sites. Excavations spread over more than one hundred years have revealed quantities of stone artifacts together with human and animal bones representing its late Ice Age inhabitants. Revised radiocarbon dating has now shed further light on the nature of the human presence here, and on the timing of the return of people to Britain after a period of Ice Age abandonment lasting about 10,000 years. Prior to this new research, it was uncertain when occupation took place and how different parts of the archaeological story fitted together, but it now seems that Gough's Cave was one of the first sites to be used by hunters of wild horses and red deer when people returned to Britain after the peak of the last glaciation.

This transformation was achieved by the dating specialist Tom Higham and the archaeologist Roger Jacobi, using ultrafiltration pretreatment on animal bones butchered or worked by the Stone Age humans, and on the remains of the humans themselves. Previously, the radiocarbon dates obtained had only made it possible to tie occupation down to a span of about 1,500 years. Now, much greater confidence can be ascribed to dates that show almost all the Upper Paleolithic material in the cave accumulated over as little as two to three human generations, centering around 14,700 years ago. Interestingly, this date corresponds precisely to a dramatic warming of climate recorded in the composition of annual layers of ice in Greenland. These archives suggest that the previously ice-covered Atlantic Ocean defrosted in about five years. Among the material dated at Gough's were bones of several of the humans, some of which show patterns of cut marks interpreted as evidence of cannibalism. Before, it had been thought that these might have belonged to a more recent phase of activity than the one associated with the horse and deer hunting, but we now know they were precisely the same age. Thus the animals, and the people who preyed on them, represented some of the first colonizers of Britain after the peak of the last Ice Age. As the climate rapidly warmed, herds of horses and deer must have migrated across Doggerland, now submerged under the North Sea, and the hunters followed.

A much older British fossil that I have been involved in studying was found in 1927 at Kent's Cavern, in southwest England. After its discovery, the anatomist Arthur Keith described this fragment of upper jaw as a modern human, but it had to wait another sixty years to achieve further fame, when it was one of the first fossil humans to be dated by the radiocarbon accelerator at Oxford. The estimated age of about 35,000 years made it one of the oldest modern humans in Europe; subsequent detective work on the Kent's Cavern archives by Roger Jacobi suggested that it could date from even earlier. So, in 2004, we decided to borrow the specimen from Torquay Museum and restudy it, using every scientific approach we could muster. The team I assembled involved researchers such as Erik Trinkaus and Tim Compton, specialists in CT and ancient DNA (techniques that I will discuss in chapters 3 and 7), curators and conservators, and Higham and Jacobi. Careful examination and CT modeling confirmed Erik's hunch that one of the teeth had been glued back into the wrong socket; a new reconstruction was made, allowing the sampling of the tooth roots for ancient DNA and ultrafiltered accelerator dating. Sadly, both of those attempts ultimately failed, but accelerator dating of animal bones found around the fossil indicates that its real age is some 40,000 years, and it may record an early spread of modern humans to western Europe.

Other physical dating methods that can be applied to fossil and archaeological materials beyond the limits of radiocarbon dating have also been developed or enhanced in the last twenty years. These include *uranium-series (U-S) dating*, which is based on the radioactive decay of different forms of uranium. Accumulation and measurement of the so-called daughter products are possible in substances like stalagmites and corals. The former has been very useful in cave sites, while the latter has been used to examine past changes in sea levels around tropical and subtropical coasts and, as mentioned already, to check the accuracy of radiocarbon measurements. One of the holy grails of dating has been to get uranium decay methods to work on fossil bones. However, this has proved notoriously difficult because, in contrast to stalagmites and corals, which are essentially sealed after deposition, bone continues to be open to the accumulation or loss of uranium (for example, as groundwater percolates through it). This means that its physical clock can run very erratically. Nevertheless considerable progress has been made recently, and I will discuss some of the results as applied to the Broken Hill fossils of *Homo heidelbergensis* in chapter 9.

A number of other methods depend on the fact that crystalline substances such as sand grains, flint, or the enamel of a tooth store up changes in electrons within their crystal structure from the radiation they receive from their surroundings, once they are buried. The amount of change (corresponding to radiation damage) can be measured from the accumulated energy released in the sand or flint when treated with a laser beam (*optically stimulated luminescence*, or *OSL*) or by heating (*thermoluminescence*, or *TL*), while in tooth enamel, the accumulated changes in the electrons can be detected using microwave radiation (*electron spin resonance*, or *ESR*). For any of these methods to work, the radiation signal must first be set at zero—for example, when a tooth begins to grow (ESR)—or set back to zero when the previous signal is wiped out as sand grains are bleached by exposure to the sun, or when flint or clay is strongly heated in a fire (luminescence). Provided the rate of subsequent accumulation of radiation damage in the material can be estimated from the environment in which it was buried, the length of time it has been in the ground (for example, in a Cro-Magnon fireplace or a Neanderthal butchery site) can be estimated.

As with radiocarbon dating, procedures have continuously been refined, so now even single grains of sand can be dated by luminescence. Equally, in the case of ESR, where previously a large chunk of a tooth had to be sacrificed, we have moved to a situation in which, using the microscopic technique of *laser ablation*, it is now possible to directly date a tiny area of fossil human tooth enamel. Another potential complication of ESR dating is the fact that fossils take up uranium when they are buried; hence they contribute to their own accumulated radiation dose. Estimating the rate of uranium uptake is critical (did most of it get in soon after burial or did it come in gradually?), but this unknown can now be addressed by combining, or "coupling," an ESR determination with a U-S date on the same piece of enamel, and looking for the age estimate that is most compatible when comparing the two.

An excellent example of the tremendous impacts that luminescence and ESR dating have made on human evolution came out of the Middle East, from the famous Israeli caves of Tabun and Skhul (Mount Carmel) discussed in chapter 1. I was fortunate enough to be involved in some of the pioneering work on dating these sites in the late 1980s and early

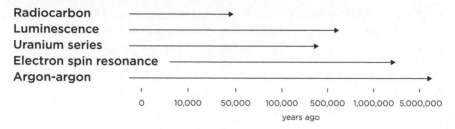

Ranges of the main dating methods for recent human evolution.

1990s, since the Natural History Museum has a share of the human fossils, artifacts, and sediments from them.

These fossils had a key role in developing ideas of Neanderthal–modern human relationships: essentially, did they represent one single rather variable population, perhaps 40,000 years old, or did the more modern-looking Skhul people succeed Neanderthals like Tabun, and perhaps evolve from them? Even when more Neanderthal fossils were added to the mix from Israeli sites like Amud and Kebara (both with quite complete skeletons from apparent burials), and more modern-looking skeletons were added from the site of Qafzeh (near Nazareth), the picture did not get any clearer. Relative dating using similarities in stone tools suggested that they were all rather close in age, while a radiocarbon date on some charcoal from Tabun suggested the Neanderthal from there was not much more than 40,000 years old. In turn, based on the known succession of Neanderthals and modern humans in Europe about 35,000 years ago, it was assumed that a similar sequence would be found in the Middle East, though perhaps it would be a little older. Thus in the early 1980s it seemed reasonable to assume that the Qafzeh and Skhul "moderns" were about 40,000 years old, and two different evolutionary scenarios were proposed in the region. Erik Trinkaus favored the view that the Neanderthals had evolved fairly rapidly into moderns there, while I took the view that there had been a replacement of the Tabun people by the Skhul and Qafzeh moderns—but we were both wrong! And there was already a clue as to why we were wrong in some relative dating work on animal remains from the sites.

Qafzeh Cave, like many of the sites, contained fossilized rodent remains as well as the human burials, and these can provide useful information not only about the local environment but also about the dating of sites. Pioneering studies of these small mammal remains by Israeli researchers suggested that Qafzeh could in fact be older than the Neanderthal sites rather than younger. This led the archaeologist Ofer Bar-Yosef to propose that the Qafzeh early moderns could date to as much as 70,000 years. Yet such an age was clearly beyond the reach of radiocarbon, so how could this view be tested? At last, with refinements in ESR and luminescence dating that came during the 1980s, this became possible.

The first significant application of these emerging techniques (thermoluminescence applied to flints that had been heated in a hearth) came from French–Israeli collaborations and initially seemed to reinforce the expected pattern in the Middle East, dating the recently discovered Neanderthal burial at Kebara to about 60,000 years. However, shortly afterward in 1988, the first application was made to the site of the Qafzeh early modern material, giving an astonishing age estimate of about 90,000 years, more than twice the generally expected figure, supporting or even exceeding the relative dating suggested by the rodents! Next up were the Skhul and Tabun sites, and for these I started working with dating specialists like Rainer Grün and Henry Schwarcz. Henry, a Canadian, is the doyen of dating in this time range, and Rainer, a German now working in Canberra, had studied and worked with him. Analyzing samples of animal teeth from both the sites for ESR dating gave equally revealing results. Within three years we had shown that the Skhul early moderns were at least as old as the Qafzeh ones, while the deep Tabun Cave sequence covered hundreds of thousands rather than tens of thousands of years. We also suggested that the Neanderthal burial from Tabun was much older than the 40,000-year radiocarbon date: it was perhaps as old as the moderns from Skhul and Qafzeh.

There was obviously a much more complex sequence than any of us had envisaged, and in some ways the expected chain of events was turned on its head: the modern-looking people from Skhul and Qafzeh were older than the Kebara Neanderthal. Further work showed that they were also older than the Amud Neanderthal. Thus they could not have evolved from these late Neanderthals, and, puzzlingly, those late Neanderthals were in the Middle East after the early modern humans, and not before them. Continuing dating work using all the available techniques now suggests that the Skhul and Qafzeh people actually range from about 90,000 to 120,000 years old, while the Tabun Neanderthal is most likely about 120,000 years old. So the emerging scenario is one where populations apparently ebbed and flowed in the region, which makes sense given its geographic position between the evolving worlds of the Neanderthals to the north and early moderns to the south.

Bar-Yosef suggested that the moderns came up into the region during a particularly warm and wet period, about 120,000 years ago, but as the succeeding Ice Age cooled and dried the north, the Neanderthals were pushed down there and took over the region—an intriguing reversal of the usual Replacement model! In fact, I think such changes could have been long-standing and even more complex, going back deep into the evolutionary history of the two species. At times when conditions were favorable, one or the other group, or perhaps both, would have moved into the region, while at times of severe aridity it might even have been completely abandoned. Whether populations were generally pushed there by unfavorable conditions in their home territories, or were pulled there by climatic ameliorations leading to population expansions, we do not yet know, but new climatic data are emerging.

The potential of ESR to match the ability of AMS radiocarbon in directly dating human fossils is at last being realized. In 1996 the first of an increasing number of applications of this technique to significant human fossils was made when Rainer Grün and I collaborated with colleagues including James Brink from South Africa to date the Florisbad human skull. This fossil, which had been found in 1932, is actually rather incomplete but seems to combine a large and fairly modern-looking face with a strong brow ridge and somewhat receding forehead. For many years it was assumed to date from about 40,000 years ago, based on a radiocarbon date from peat deposits at the site, and on that basis it seemed to be a relic hanging on in the margins of southern Africa, while moderns were evolving and spreading in western Asia and Europe. As such, it supposedly demonstrated the backward role of Africa in modern human evolution—the primitive Florisbad humans merely marking time until moderns arrived from farther north and replaced them. However, the fossil preserved one upper molar tooth, and a tiny fragment of its enamel was taken to Rainer's lab in Australia for ESR dating—with sensational results. The fossil was not 40,000 but about 260,000 years old! Thus its potential role in human evolution was revolutionized at a stroke: rather than representing a southern African equivalent of the Neanderthals, on the brink of extinction, it could instead have been an ancestor to us all.

There are some situations where even the best physical dating techniques need a helping hand, and combinations of physical and relative methods are required. The Neanderthals apparently disappeared about 30,000 years ago, but the factors leading to this and the time scale for their demise are still fiercely debated. While accelerator radiocarbon dating gives excellent precision in the measurement of an age, it does have problems of accuracy compared with calendar years during this critical period of time, both because the rate of formation of radiocarbon in the atmosphere was unusually variable then, and because even a tiny amount of contamination from young or old carbon will make a significant difference to the age obtained. As I explained earlier, this latter problem is being addressed through techniques that very effectively remove contaminants before dating is attempted. But to address the former issue, there were fortunately other significant events in Europe during that period to provide new and potentially very accurate ways of relative dating. As I discuss further in chapter 4, a massive volcanic eruption took place in the Campania region of central Italy about 39,300 years ago (which we know from argon dating). As well as enormous quantities of local deposits such as lava, pumice, and ash, the eruption also produced much finer volcanic dust, known as *crypto-* or *microtephra* because it cannot be seen with the naked eye. This microtephra may be ejected into the upper atmosphere and travel for many thousands of miles, and the Campanian Ignimbrite—from the Latin words *igni* (fire) and *imbri* (rain)—settled eastward as far as Russia and North Africa.

The CI, as it is known, has now been found in dozens of archaeological sites, including the famous Russian localities at Kostenki, in levels which we already knew from radiocarbon dating were at least 35,000 years old. Each volcanic eruption took place as a result of unique combinations of factors like chemical composition, temperature, and pressure, and thus can be "fingerprinted" and recognized. So wherever the special CI chemical signature is found in an archaeological site, we can be pretty confident that the level concerned, with its associated fossils and artifacts, was laid down just over 39,000 years ago. In turn, all such sites can be correlated to this age by a lattice of synchronous volcanic deposits.

This approach gave rise to a large collaborative project called RESET (Response of Humans to Abrupt Environmental Transitions), in which I am involved. Over a five-year research period, RESET is correlating tephras from their volcanic sources to where they fell in deep ocean and lake sediments, and even farther into important archaeological sites in Europe, western Asia, and North Africa. The aim of RESET is to investigate the effects of climate and environmental changes on the human populations of the region, including the last Neanderthals and the first moderns. The tephras themselves are markers of volcanic eruptions, of course, most of which were only local and short-lived in their effects, but a few did have major—even global—impacts, as we will see later in this chapter.

Using volcanic deposits to date human fossils has a long history, as I explained in relation to Olduvai Gorge earlier, and the mapping of outputs from successive volcanic eruptions has played an important role in refining the age of many important fossil sites in East Africa, including Omo Kibish in Ethiopia. The two most complete human fossils from there, the Omo 1 skeleton and the Omo 2 braincase, were found in 1967 by a team led by Richard Leakey and were important in early proposals for a recent African origin. But although there were initial estimates that the material was over 100,000 years old, some of these were based on the application of uranium-series dating to shells in the deposits—not the most reliable material for such determinations—and so doubts remained. Over thirty years after the original discoveries, an international team led by the anthropologist John Fleagle returned to the Kibish region, relocated the 1967 find-spots, and found further fossils and stone tools. Both Omo 1 and Omo 2 were originally recovered from the lowermost portion of the massive Kibish Formation, a series of annual but episodic sediments laid down by the ancient Omo River when it periodically flooded, before it entered Lake Turkana. These deposits lie about one hundred kilometers farther north than its present delta, close to the Ethiopian border with Kenya. Occasionally, volcanic eruptions deposited volcanic ash and pumice over the river and lake sediments, and these can be dated through their contained argon. A layer of ash about three meters below the location of Omo 1 was placed at about 196,000 years old, while a second ash about fifty meters above the location was dated to about 104,000 years. Because there were also clear signs of geologic erosion (the removal of sediments when the river and lake level fell) between the level of Omo 1 and the higher ash, it seemed likely that the age of Omo 1 was much closer to the age of the 196,000-year ash than to the 104,000-year one.

Additional indirect support for this came from much farther afield, on the seabed of the Mediterranean. During ancient monsoon periods, rain and snowmelt in the Ethiopian highlands sent annual floods pouring into the sources of the River Nile, causing *sapropels* (dark layers of sediment) to be deposited when these waters eventually flowed out into the Mediterranean. A particularly strongly marked sapropel can be dated from its position in Mediterranean seabed cores to about 195,000 years, suggesting that it correlates perfectly with the major monsoon event that sent floods in the opposite direction down the Omo River, producing the vast deposits of the lower part of the Kibish Formation, in which the Omo 1 skeleton and the underlying volcanic ash were found. The Omo 2 braincase was a surface find rather than a fossil excavated from sediments (which was the case for the Omo 1 partial skeleton), but the surrounding location consisted of the lowermost part of the Kibish Formation. Thus the team that revisited the region and published the new dating work remains confident that Omo 1 and Omo 2 are very close in age, at about 195,000 years, despite some strong contrasts in their level of modernity—something to which I will return in chapter 9.

Another instance in which Mediterranean sapropels provided clues about events deep within the African continent concerns the "greening" of the Sahara about 120,000 years ago. Today, the Sahara is the largest hyperarid region on Earth, with an annual recorded rainfall as low as one millimeter across much of its vast extent. But as is well known from archaeological finds and rock paintings of animals and people deep within the desert, only 6,000 years ago the Sahara was a wetter place of grasslands, lakes, and gallery forests, fringing extensive river systems. What is less well known is that 120,000 years ago the Sahara was even wetter than that and was able to support a widespread population of Middle Paleolithic hunters and gatherers. There are many clues to this, including chemical signs of freshwater deposition and dark sapropels in deep-sea cores off the coast of Libya, both of which indicate powerful river sediments during the last interglacial. A similar signal was detected from plant-derived chemicals in dust deposits off the West African coast; analysis shows that the level of water-dependent plants (such as trees rather than grasses) peaked around 115,000 years ago, with a briefer second peak from about 50,000 years.

In addition, as revealed in radar images from satellites, huge river channels now lie buried beneath desert sands, some of which are five kilometers wide and run for eight hundred kilometers. As the earth scientist Nick Drake and his colleagues showed, around 120,000 years ago the desert was covered by an interconnecting network of rivers and lakes, forming humid corridors that stretched from massive southern lakes such as Fazzan and Chad all the way to the Mediterranean. These corridors allowed typical African fauna and flora to flourish for at least 20,000 years, and as plant resources and game proliferated, so did the humans who lived off them. For the last hundred years, travelers and archaeologists have collected Middle Paleolithic tools from the surface of the Sahara, often well away from any modern oases, and we now know that many of those accumulations date from the green Sahara over 100,000 years ago.

The tools include triangular stone points with a tang or shoulder, which was presumably used to mount them as projectiles on a wooden handle. These iconic artifacts characterize the Aterian industry, first recognized at the Algerian site of Bir el-Ater; this industry was made by a very robust and large-toothed variety of early *Homo sapiens*, comparable to the people we know from Herto in Ethiopia. It seems very likely that the increased humidity of many parts of Africa at this time led to population growth and an important sharing of ideas across Africa, as formerly isolated regions became connected by habitable corridors. Cultures that used shell beads and red ocher pigment to signal to each other seem to have spread across the whole known range of early modern humans at that time, from South Africa to Morocco and even into adjoining western Asia, at Skhul and Qafzeh. In Israel, elevated rainfalls produced the huge Lake Samra, which extended far beyond the now shrunken Dead Sea basin, about 75,000 to 135,000 years ago.

But as the beginnings of the last Ice Age began to grip, these balmy interglacial conditions were not to last in Africa. We can trace the effect of this climatic downturn on the peoples of southern Africa from about 75,000 years ago through two important and innovative stone tool industries: those at Still Bay and Howiesons Poort. As well as sophisticated stone artifacts (heat-treated to improve flaking qualities in the case of the Still Bay), both possessed beads made of seashells or ostrich eggshells, and both used red ocher symbolically. The Still Bay industry is known from only a handful of spots in southern Africa, while the Howiesons Poort was much more widespread, with at least thirty sites ranging from the best-known locations on the southern coasts at places like Klasies River Mouth Caves, to the edges of the Namib Desert and the mountains of Lesotho. While it was thought that the Still Bay preceded the Howiesons Poort, they both lay well beyond the limit of effective radiocarbon dating, so methods like uranium series and ESR had been employed to place them in relation to each other, but with a poor fix on their respective durations.

The breakthrough came in 2008 when a team of dating specialists including Zenobia Jacobs and Bert Roberts combined with archaeologists such as Hilary Deacon and Lyn Wadley to apply the latest luminescence dating techniques to single grains of quartz from the sites, using the same laboratory procedures throughout. Fifty-four samples were obtained from widely dispersed sites that contained either both industries or one or the other. The results were striking: rather than spanning periods of 50,000 years or more, as some other methods had suggested, both the Still Bay and the Howiesons Poort industries were rather brief cultural episodes that seemingly appeared and vanished quite suddenly over large areas of southern Africa. Remarkably, the Still Bay lasted only a few millennia, around 72,000 years ago, while the Howiesons Poort appeared around 65,000 years ago, ending abruptly at about 60,000 years. Moreover, the people who succeeded the Howiesons Poort only returned after a gap of a few thousand years and were making more conservative Middle Stone

Age tools (comparable to the Middle Paleolithic of western Eurasia), apparently without the innovations of their predecessors.

It is, of course, possible that the manufacturers of the previous industries did not disappear but simply moved to locations that have not so far provided any archaeological records. (For example, they might have relocated farther out on the coastal shelves, which are now submerged.) But they do not seem to have reappeared even at a later date, suggesting that these really were brief episodes, like a light turning on and then being extinguished, perhaps forever. Environmental deterioration in the face of rapid climate change has been invoked to explain these episodic patterns, and I will return to their significance in chapter 8. Meanwhile, I would like to examine a global event that has controversially been claimed to lie behind even wider changes in human populations and behaviors, including the innovations of the Still Bay industry: the eruption of the Toba volcano in Sumatra.

About 73,000 years ago, the large island of Sumatra in Indonesia was the source of the most powerful volcanic eruption of the last 100,000 (some calculations suggest 2 million) years. The eruption was about a thousand times larger than the famous Mount St. Helens in Washington State in 1980, and it expelled the equivalent of about 1,000 cubic kilometers of rock in the form of ejecta of many different sizes, as well as huge volumes of water vapor and gases. Thick ash deposits from the eruption were found in cores stretching from the Arabian to the South China seas, and some archaeological sequences in India are interrupted by ash falls several meters thick. The undoubted scale of the eruption led to some sensational claims about its effects on the Earth through a resultant "volcanic winter," where the whole planet would have lacked summers for many years, as a result of clouds of dust and droplets of sulphuric acid residing in the upper atmosphere. The resultant drop in temperature and the lack of summer sun would have devastated plant growth and everything that relied on it, including the early human populations of the time. Some suggested it destabilized the Earth's climate for a thousand years, or that it even triggered a global ice age, shrinking human numbers to only a few thousand people. On the other hand, studies of faunas in southeast Asia, closest to the eruption, suggested that any effect was minor and short-lived, since they were not devastated. Moreover in India, archaeological sequences studied by Mike Petraglia and his colleagues similarly indicate that the impact on human populations was not severe. I have been very cautious about Toba's effects on humans globally. (After all, the Neanderthals living in temperate Europe and the Hobbit living in Flores in Indonesia, as well as our ancestors in Africa, certainly survived the effects of Toba somehow.)

However, two recent studies by Alan Robock and his colleagues and Claudia Timmreck and her colleagues, using different models of its effects around the globe, do point to a severe, if shorter-lived, impact. Their work did not back up the idea that it could have triggered a glacial advance, but did conclude that it could have produced up to a decade of cold, dry, and dark conditions, serious enough to affect plant and animal life on land and sea, but perhaps not totally devastating. On the other hand, new analyses of land sediments and pollen in a core from the Bay of Bengal, involving the leading proponent of the Toba effect on early humans, Stanley Ambrose, did find signs of a long period of desiccation in India at the time of the Toba eruption. Unfortunately for simple scenarios, the apparent extinction of the South African Still Bay people after a short florescence came about 2,000 years after Toba, if the present dating evidence is accurate, although some might argue that their innovations were an outcome forced by the environmental degradation that the eruption brought about.

Now let's move on about 35,000 years to the environments and time scales around the time of the extinction of the last Neanderthals. Of course, if the Neanderthals passed on their genes to modern humans, as we will discuss in chapter 7, they did not become entirely extinct, since some of their DNA lives on within us. Nevertheless, as a population with their own distinctive bodily characteristics they vanished, and there are many scenarios constructed around what may have happened to them. Explanations have ranged widely from suggestions of imported diseases to which they had little natural immunity, through to economic competition from, or even conflict with, early modern humans. Until recently the view of climate change in Europe at this time was rather simplistic, leading to climate being ignored as a factor in Neanderthal extinction: they became extinct before the peak of the last Ice Age, they had survived cold conditions before, and they were adapted physically, and probably culturally, to cope with climatic downturns. Many explanations (including my own) had instead focused on the direct impact of modern humans on the Neanderthals, and the inherent superiority of people like the Cro-Magnons. But rich paleoclimatic records from cores in ice caps, the sea floor, and lake beds now reveal a startling complexity in climatic change at this time, with many rapid oscillations. This led to new ideas about their extinction, including those of two of my friends, Clive Finlayson and John Stewart, who consider that the Neanderthals were probably on the way to extinction anyway, and that moderns had little or nothing to do with their demise. For example, Clive thinks that early moderns had honed their adaptations on the plains of Africa, a very different environment from that to which the Neanderthals were adapted in Europe; hence the two species had different ecological preferences and never really overlapped, competed with each other, or interbred. According to this view, the Neanderthals faded away about 30,000 years ago as their preferred mixed habitats finally vanished from their last outposts, in places like Gibraltar.

In my own case, around the year 2000, I took part in a collaboration called the Stage 3 Project (Marine Isotope Stage 3 lasted from about 30,000 to 60,000 years ago), led by Tjeerd van Andel and based in Cambridge. We used fluctuations in temperature recorded in a Greenland ice core and a lake core in central Italy to reconstruct hypothetical "stress curves" for Europe, based on two factors of equal weight, both of which were assumed to be bad for humans, whether Neanderthal or Cro-Magnon: low temperatures and rapid destabilizing fluctuations in temperatures in either direction—higher or lower. The approach was simplistic in that it did not attempt to model other factors such as changes in rainfall, snow, and wind chill, which would also have had important impacts on the human population of Europe and their survival prospects. The stress curves we generated showed a mild climatic phase around 45,000 years ago, which perhaps correlated with the migration of moderns into Europe, but the "climatic stress" peaked at around 30,000 years, rather than the subsequent glacial maximum, coinciding with the last known records of the

Neanderthals or their stone tool industries in places like Gibraltar and the Crimea. Such stressful conditions would no doubt have affected both Neanderthal and Cro-Magnon populations, sharpening the competition for diminishing resources in environments where their ranges overlapped. But only *Homo sapiens* came through those crises.

A rather more sophisticated modeling of conditions in Europe when the last Neanderthals overlapped with the Cro-Magnons was published in 2008. In this work William Banks and his colleagues used the location of stone tool assemblages dated between about 37,000 and 42,000 years, which were thought to identify the presence of Neanderthals or Cro-Magnons in particular regions. Then, treating the distributions as though they represented a species of mammal rather than stone tools, they used ecological modeling to reconstruct the environmental preferences and tolerances of the two populations, and the ranges they each should have been able to occupy at the time, according to those preferences. The time span covered two mild phases interrupted by a short but severe cold snap at about 39,000 years. This was not when the Campanian Ignimbrite was deposited farther to the east, but when the Atlantic was chilled for several hundred years by the southward flow of an armada of icebergs (a *Heinrich event*, discussed further in chapter 4).

The results showed that, before the cold snap, the Neanderthals should have been widely distributed, and indeed they were. During the Heinrich event, both populations shrank in their modeled and actual distributions, in the face of the environmental deterioration. But when conditions ameliorated at about 38,000 years, although the warmer and wetter conditions should have encouraged both populations, the moderns bounced back while the Neanderthals did not. The modeling also showed that the Neanderthal and Cro-Magnon populations were attempting to exploit similar ecological niches; in practice the moderns increased the breadth of theirs through time, at the expense of the Neanderthals. While in the earlier phases the modern niche did not include central and southern Iberia, as time went on the Cro-Magnons increasingly expanded southward toward outposts of Neanderthal survival such as Gibraltar.

This interesting work shows how such modeling can be done, and it should be possible to further refine the analyses, as ultrafiltered radiocarbon dates and the use of correlation tools like microtephra become increasingly available. A more direct attempt to estimate the relative population sizes of the last Neanderthals and the first moderns in western Europe came from the Cambridge archaeologists Paul Mellars and Jennifer French. They mined a large data set recording the extent in area of each of the last Neanderthal sites in southwestern France and those of the succeeding Aurignacians in the same region. Similarly they compared data on the number of stone tools each human population left behind in their sites, and the amount of food debris they generated. Multiplying all these together, they concluded that the early modern population was about ten times the size of the preceding Neanderthal one. This might imply that the moderns swamped the Neanderthals, but at the moment we can't reliably place them together as direct competitors in the European landscape over any precise length of time, only infer that they probably did coexist.

As I have come to realize, we should not be looking for a single cause for Neanderthal extinction anyway; we need to take a wide view of this. The fascinating events that took place in western Europe some 35,000 years ago get most of the scientific and popular attention, but they were only the endpoints of hundreds of thousands of years of evolution and potential interaction between the lineages of modern humans and the Neanderthals (for example, ancestral populations could have been in contact intermittently in regions like western Asia). I am sure there were differences (many unknown to us) in appearance, communication, expression, and general behavior that would have impinged on how Neanderthals and moderns saw each other. So when the populations met, did they perceive each other as simply other people, enemies, alien, or even prey? And since the Neanderthals disappeared at different times across Asia and Europe, the reasons why they disappeared from Siberia might be different from why they became extinct in the Middle East, and again different from the factors at work in Gibraltar or Britain—and these factors may not always have included the presence of modern humans.

This brings us back to one of the favored explanations for the extinction of people like the Neanderthals: behavior. I am one of those who have often invoked the behavioral superiority of modern humans over other human species as the main reason for our success and their failure, but reconstructing such behavior from the archaeological record, let alone deciding who is superior to whom, is no easy matter. In the next chapter we will look at new methods of unlocking evolutionary and behavioral insights from the fossils, and then in the succeeding two chapters we will consider what the archaeological record now seems to be telling us.

3

What Lies Beneath

THE FOSSIL RECORD OF THE EARLY HISTORY OF OUR SPECIES AND our close relatives such as the Neanderthals has grown tremendously in the last twenty-five years. But what has developed at an even faster rate is our ability to unlock secrets from those fossils, secrets that tell us about the biology and the lives of those long-dead people. In this chapter I will show how new techniques let us look at the size and shape of ancient skulls, and reveal hidden structures such as the inner ear bones that can tell us about the posture, movement, and senses of vanished peoples. Now we can look at butchery marks microscopically to examine details of ancient human behavior, daily growth lines in fossil teeth to reconstruct how children grew up 1 million years ago, and we can use isotopes to reveal how ancient humans in different parts of the world exploited their environments, and what they ate. In the last twenty years traditional methods of recording the size and shape of fossil bones and teeth have been complemented and increasingly superseded by techniques that capture such information on a computer, through digitizing or scanning. The medical technology of *computerized tomography* (*CT*) X-raying has been particularly successful in extending work into anatomical structures that either are difficult to measure through traditional techniques (for example, the shape of a curved form like a brow ridge) or are otherwise inaccessible (for example, fossils hidden inside rocks or unerupted teeth in a jawbone). And the computational technique of *geometric morphometrics*

(*morphometrics* simply means "measuring shape or form") is allowing wider and more detailed comparisons of the size, shape, and even the growth patterns of fossil and recent samples.

Most of these new techniques were not available when I began my research on human evolution, and they were still in their infancy when Recent African Origin models started to germinate in the 1980s. For example, when I made my four-month trip around Europe to measure about a hundred fossil skulls of archaic and modern humans in 1971, I carried a small suitcase full of metal measuring instruments, such as callipers, tapes, and protractors, and a camera to record the preservation and basic shape of the specimens I was studying. (A single well-preserved skull with a lower jaw might take up to half a day to record fully.) With no laptops or pocket calculators, all my data were slowly recorded by hand on paper sheets, and, without photocopiers, there was a great risk (unappreciated by me at first) that all the hard-won data on which my career depended could easily have been inadvertently lost or even stolen in the two thefts that I suffered from my car.

When I returned to Bristol, I took months to laboriously transcribe my data onto punch cards and start the computational wheels turning on the single, massive (but in modern terms ridiculously underpowered) computer that served the whole of Bristol University. Now, a single researcher could (if he or she knew where to look) achieve the equivalent amount of data collection, which took me four months and about 5,000 miles of travel, in a few days sitting at a computer console, summoning up online measurements and CT scans. Far more sophisticated comparative analyses of skull shape than I achieved in two further years could probably be accomplished in a few more days! Still, I have no doubt that by directly studying the fossils, I did obtain insights into their nature that would not have been apparent had I been sitting at a remote workstation. Plus I had the thrill and honor of holding and studying firsthand such iconic fossils as the skulls from the Neander Valley and the Cro-Magnon rock shelter.

The approach I used to measure and compare my samples of fossil and recent human skulls is now called *conventional morphometrics*, although in 1971 this was very much the standard approach and had been in use since before the time of Charles Darwin. The human skull has various points where bones meet, where muscle markings cross a bone, or where there are specific locations such as the external earhole or the widest breadth of the nasal opening. These "landmarks" are used as measuring points so that an instrument can, for example, be laid across the nose to record its breadth at its widest point, or can measure the total length of the braincase from the top of the nasal bones at the front to the farthest point away in the middle of the occipital bone at the rear. Measurements and their variations can then be directly compared between specimens, either singly or through the calculation of an index or angle, using two or more of the measurements. For example, the *cephalic index* (*CI*) was a much used ratio of the breadth of a skull to its length. This index was a basic measure of how long or broad-headed a skull was, and in some of the racist science of the last two centuries it was taken as a crude measure of "primitiveness," on the assumption that the most backward "races" had the longest heads.

My Ph.D. work made use of angles and indexes but also extended to the then relatively novel area of *multivariate analysis*, where a large number of measurements could be assessed together, with specimens compared in a computed space of many dimensions, or via a single *distance statistic*—a bit like a ratio or index but one calculated from many measurements combined, rather than just two. However, I realized even then that my measurements were not capturing the whole complex shape of a skull, particularly some of its curved surfaces, which were poorly marked by suitable landmarks. And it was evident when comparing small and large skulls even within a single population that they might change their relative proportions as they changed in size (the study of this is known as *allometry*), which was difficult to capture and visualize effectively with the techniques I had available in the 1970s.

Today a new approach called *geometric morphometrics* allows far more effective visualization and manipulation of the shape of a complex three-dimensional object such as a skull. The whole shape is captured through scanning or digitizing, and virtual landmarks can be created by software at intervals across the surface of the object in question, such as a skull or jawbone. However, these secondary landmarks are still usually anchored to a network of primary points that correspond between the different objects to be compared, to provide a common frame of reference. A grid of points that reflect the overall shape (for example, of a Cro-Magnon skull) can be displayed on a screen, and a similar grid from another object (for example, a Neanderthal skull) can be compared side by side or overlaid. Geometric morphometric software can then reduce the skulls to the same overall size and measure the amount of shape distortion needed to change one to the other, and also quantify which areas of the skull change the least or the most in such comparisons. Thus research by anthropologists like Katerina Harvati has provided strong evidence that the shape differences between modern and Neanderthal skulls is certainly at the level of species differences in recent primates. These techniques can also be used to show how a series of skulls change as they mature and can even create theoretical evolutionary intermediates—say between a *Homo erectus* and a modern skull, which can then be compared with real examples, such as the skull of a *Homo heidelbergensis*. Geometric morphometrics has become particularly important when used in conjunction with that other technology recently applied to fossils: computerized tomography (CT).

Wilhelm Röntgen, a German physicist, is usually credited with the discovery of X-rays, in 1895. He chose the name to reflect their then unknown nature, and his serendipitous find was to win him a Nobel Prize. He recognized their possible medical uses after he experimented by "photographing" his wife's hand with the new discovery, clearly revealing the bones inside, and the technique was soon applied to new discoveries of fossils, such as the Neanderthals from Krapina in Croatia and the *Homo heidelbergensis* jawbone from Germany. X-rays found many uses in paleoanthropology over the succeeding century, but the flat form of the conventional X-ray image meant that structures could obscure each other, and they were not all correctly scaled in relation to each other (just as a shadow can be out of proportion to the original shape of an object).

Not long after Röntgen's discovery, an Italian radiologist, Alessandro Vallebona, proposed a method that provided a more focused single slice of X-ray imagery, which became known as *tomography* (from the Greek words *tomos* [slice] and *graphein* [to write]). This method found many uses in medicine, and about forty years ago Godfrey Hounsfield in the United Kingdom and Allan Cormack, working in the United States, independently came up with the development known as computerized tomography (CT), for which they also jointly won Nobel Prizes. CT scanners send several beams simultaneously from different angles, after which the relative strengths are measured, and a two-dimensional slice, or a three-dimensional whole structure, can be reconstructed from the data received. The computerized images reflect the density of the tissues or materials through which the beams have passed; for example, air spaces let through a strong signal, and teeth or fossil bone a much weaker signal. Moreover, the ability of CT to provide focused images means that much more detail is available than in conventional X-rays—so even the microstructures of bones and teeth can be examined.

As CT technology and computing power have escalated, so has the impact on studies of human evolution. In some of the earliest applications in the 1980s, Javanese *Homo erectus* fossils were CT scanned, showing previously unseen inner ear structures, but the quality of the images was not really good enough to reveal their evolutionary patterns. But within ten years, things had moved on so far—in research pioneered by the paleoanthropologist Fred Spoor—that it was possible to image and compare the minute inner ear bones of several Neanderthal fossils, showing for the first time that they were distinct in shape from those of modern humans.

Anatomically, our ears are divided into three parts: outer, middle, and inner. The outer ear gathers and transmits sound waves via the eardrum into the middle ear, where its tiny chain of bones converts the energy into mechanical vibrations. These middle ear bones—the malleus, incus, and stapes—are sometimes found in or next to the ear canal of fossil skulls, and so they have been studied in a few cases without the necessity of CT. Thus we know that early Neanderthal fossils from the Sima de los Huesos site at Atapuerca in Spain had middle ear bones shaped like ours, with the implication that the perception of sound in these early Neanderthals was already similar to our own today. The vibrations transferred via the middle ear bones then pass through the fluid and membranes of the cochlea of the inner ear to become nerve impulses, which are finally transmitted to the brain so that we can perceive the sounds. But our ears are not just for hearing; two other parts of the labyrinth of the inner ear also help us control our balance and head movement. The first consists of two fluid-filled chambers, which are lined by small hairs that sense the movement of tiny crystals of calcite, so that we can balance our head properly. The second consists of three fluid-filled loops that are arranged at ninety degrees to each other. These semicircular canals are also lined by hairs, which, through the motion of fluids on them, sense the movement and rotation of the head, and it is these canals that have proved so interesting when comparing the Neanderthals and all other humans. In particular we know that the size and shape of the semicircular canals are laid down before birth and remain unchanged as we grow up, and thus any differences that exist are likely to be genetic in origin and are largely unaffected by the environment during life.

Nearly twenty Neanderthals have now been CT scanned to reveal their inner ear anatomy, and each semicircular canal is subtly distinct in size, shape, and orientation when compared with ours. What makes this discovery particularly intriguing is that the canals in the assumed ancestral species *Homo erectus*, and in early modern human fossils studied so far, are more like our own, so it seems to be the Neanderthals that are the odd ones out. But fossils in Europe that might be ancestors of the Neanderthals, such as the skulls from Steinheim and Reilingen in Germany, show an approach to the Neanderthal conformation, suggesting that the distinctive pattern could have evolved in Europe. But why?

One possibility is that the shape of the semicircular canals is reflecting something else, such as overall skull or brain form, and it's true that the Neanderthals do have some distinctive features in the shape of their temporal bones—the bone around the ear region on each side of the skull. Another possibility is that it reflects some form of adaptation—perhaps climatic—but against this, modern humans from cold climates show no significant differences from moderns who live in hot climates. The scientists who have conducted the most comprehensive studies, including Fred Spoor, argue that a plausible explanation lies in the essential function of the semicircular canals: to control the movement and rotation of the head. Although the exact mechanisms of interplay between the head and neck and the semicircular canal system are poorly understood, the Neanderthals had shorter but bulkier neck proportions than modern humans, which might have affected movements of the head, if it was more deeply buried within powerful shoulder and neck muscles. In addition, the Neanderthals had a more projecting back to the skull, a flatter base to the braincase, and a more projecting face, particularly around the nose, all of which might have made a difference to head movements all the way from less strenuous activities like walking through to highly energetic running or hunting.

One of the first fossils to reveal these unusual Neanderthal inner ears forms part of the collections at the Natural History Museum in London. This is the rather fragmentary Devil's Tower child's skull (from its large size, probably a boy), found together with animal bones and stone tools beneath the sheer north face of the Rock of Gibraltar during excavations in 1926. It consists of three skull bones, half of an upper jaw, and most of a lower jaw, with a mixture of milk teeth and still-forming permanent teeth. In modern children, where the birth date is uncertain, or an unknown murder victim needs to be identified through forensics, the best way to estimate age is from their teeth. This method was applied to the Gibraltar fossil, and it was straightaway evident this was a child less than six years old in modern terms, as the first molar tooth was not yet ready to erupt. Studies of the fossil in 1928 suggested from the dental maturity that he was actually about five years old at death, but to judge from the voluminous skull bones, his brain size was already slightly larger than the modern average. Until 1982, everyone assumed that the bones belonged to one child, but in that year the anthropologist Anne-Marie Tillier suggested that while most of the bones did indeed represent a child of about five, the temporal bone was from a different and less mature child of about three years old at death.

In the 1970s new microscopic techniques became available to study the microstructure of teeth, and earlier suggestions that human tooth enamel contained daily "lines" of growth began to be studied as a means of estimating the length of time it took a tooth to develop, and hence of potentially gauging the age at death of a child. These daily lines are grouped and expressed on the surface of the front teeth as transverse ridges, or *perikymata* (from two Greek words meaning "around" and "a wave"), each of which represents about eight days of growth. In the 1980s, using a scanning electron microscope, I collaborated with the paleoanthropologists Tim Bromage and Christopher Dean, and subsequently also with the primatologist Bob Martin, to estimate the probable age of the Devil's Tower child from its well-preserved upper central incisor, and to study its growth and development. By counting the perikymata and adding a few months to represent the small amount of root growth, we estimated the age at about four years. We also used a rare and important collection of human skeletons from the crypt of Christ Church at Spital-fields in the City of London, with actual age at death recorded on coffin plates or parish records, to test the perikymata method. We found that it worked well as an estimator of age in the children who had been buried there. In addition, I studied the temporal bones of the same children's skulls to assess whether a temporal bone as immature as the one from Devil's Tower could belong with the other bones and teeth of that child. The results were clear: both the teeth in the jaws and the temporal bone came from a child of about four years at death, and thus there was no reason to dissociate them on grounds of differential maturity. However, because the temporal bone was from the other side of the skull to the equivalent parietal bone, the two could not actually be directly articu-lated to prove they belonged together.

Nevertheless that "fit" was demonstrated a few years later when the CT experts Christoph Zollikofer and Marcia Ponce de León used the tech-nique to reveal further anatomical data and to produce a three-dimensional reconstruction of the whole skull, showing that the temporal bone undoubt-edly belonged with the other remains. They not only mirror-imaged the missing parts from the preserved portions but were also able to complete a hypothetical whole skull by "importing" elements from other Neanderthal children who had the appropriate parts preserved, adjusting their size vir-tually to complete the fit. To test the method, the researchers also virtually disarticulated a modern child's skull of comparable maturity and demon-strated that they could re-create it very accurately, using only the portions preserved in the Gibraltar child.

Having re-created the Devil's Tower child's skull digitally and on-screen, they could also remake it physically, using a technique called *ste-reolithography*. This technique was developed for industrial purposes to test the fit of parts with each other, and rather than carving out or molding a shape, objects are built up via the consecutive solidification of thin layers of a light-sensitive liquid resin. It is magical to watch the process—an ultraviolet laser beam, guided by the digital CT data, gradually material-izing a solid object out of a pool of transparent resin. A skull or jaw can be re-created, one thin layer at a time, as the beam flickers across the resin, causing the liquid to progressively set. This replication method has many advantages over conventional molding and casting: it causes no damage to the surface of valuable fossils since it is noninvasive, it is remarkably accurate and nondistorting, and internal structures such as air spaces and unerupted teeth can be replicated and made visible if the transparent resin is left uncolored.

But this was not all that was revealed. The boy's teeth (including those still unerupted in the jaws) were also studied in great detail, and a feature that had been noted in previous research was given special attention. The front teeth in the two halves of a lower jaw are usually mirror images of each other in terms of their positions and orientations, but in the Devil's Tower specimen, some on the right side seemed out of place. The CT images clearly showed that this boy had suffered a fracture of the lower jaw earlier in life but had survived, allowing the injury to heal quite well, and so this was unlikely to have been the cause of his early death. As already mentioned, he was large-brained, and the CT reconstruction also allowed an accurate estimate of his brain size, which would have been between 1,370 and 1,420 cubic centimeters, with a little more growth to come—a volume already comparable with those of European men of today.

There has been much discussion about how Neanderthals grew up—whether they matured at the same pace as we do today—and the Devil's Tower child has become an important part of the discussion. Apes have rapid brain growth before birth and relatively slower growth in the years immediately after, while we have rapid brain growth both before and after birth. At birth, allowing for body size differences, human babies already have brains that are one third larger, relatively, than those of apes, but by adulthood our brains are three times larger. The fact that we must grow our brains so much after birth is largely dictated by the limits imposed by the size and shape of the birth canal of the human pelvis, and it's likely that there is a limiting threshold of about five hundred cubic centimeters, after which a substantial period of postbirth growth in brain volume would be required.

This threshold must have been reached during the time of *Homo erectus*, which means that *erectus* babies probably had extended periods of immaturity compared with apes, during which the brain could continue to grow at a fast rate. For example, estimates suggest that compared with our landmarks for the average eruption age for the first, second, and third molars of about six, twelve, and eighteen years, *erectus* may have had a timing of about five, nine, and fifteen years respectively. But that eruption sequence marking important stages in childhood, adolescence, and the beginning of adulthood would still have been far more prolonged than in the chimpanzee, whose molar eruption ages are about three, six, and ten years.

Essentially apes have an infancy of about five years, after which they have about seven years of adolescence and are then projected into adulthood, whereas modern humans have two extra phases inserted between infancy and adolescence: childhood (about three to seven years) and a juvenile phase (between about seven and ten years). In these phases the child is still dependent on support from its mother and older kin, for protection, for learning, and for food to grow and fuel an energetically demanding large brain. The fact that our children grow so slowly spreads out the energetic costs of rearing them and may be an important contributing factor in the greater number of children that *Homo sapiens* parents can sustain, compared with apes. And recent studies have shown that although adult human brain size is essentially achieved by the age of eight, the brain continues to wire up its connections and cross-connections right through adolescence, when in humans there is still much to learn culturally and socially. In addition, we mature much later than the other apes, with an adolescent phase lasting between the ages of about ten and eighteen. Neanderthals with their large brains must have had long childhoods too, although, as we shall see, there is some evidence that they reached adulthood earlier than the average for humans today—not surprising, and perhaps even essential, if most adults were likely to die before they reached forty (see chapter 6). So their learning processes would have been extended too, even if not quite to the extent we find in our species, and their brains may have had to grow to their large size at a slightly faster rate and over a shorter period—which perhaps explains some aspects of their diet. Whether their large brains endowed them with an intelligence like ours is another fascinating question.

The brain and head size of *Homo sapiens* at birth are right at the limit of what is practicable for the human birth canal to withstand, and medical science may be required to assist in difficult deliveries, taking over the role of midwives in traditional societies. There are a few poignant

Cro-Magnon burials of women with seemingly newborn babies, attesting to the difficulties of birth 30,000 years ago. A notebook account of the 1932 excavation of the much more ancient Tabun Neanderthal woman's burial, in what was then Palestine, mentions the skeleton of a fetus tucked against the side of her body. Sadly, these enigmatic remains were never described, and we do not know whether this was a mistaken identification or if the material was too fragile to recover from the hard cave sediments. But the woman's skeleton did survive and is curated at the Natural History Museum, representing the most complete female Neanderthal skeleton yet described (others from the Sima de las Palomas in Spain are in the process of study and publication by Erik Trinkaus and colleagues).

The CT experts who worked on the reconstruction of the Devil's Tower child's skull also worked on reconstructing the pelvis of the Tabun woman. In the absence of the putative remains of the baby that was found with it, they instead reconstructed the fragile skeleton of a newborn Neanderthal child buried at Mezmaiskaya in Crimea, and in a spectacular demonstration of the power of CT technology, they combined the two in order to investigate Neanderthal obstetrics. They discovered that the child's brain size was about four hundred cubic centimeters, typical of a newborn today, but the skeleton was already much more strongly built. In testing the birth process, it was apparent that the slightly wider pelvis of the Tabun woman should have eased labor. However, the baby's skull was already Neanderthal-shaped, with a longer head and a more projecting face, suggesting that the birth process would have been as difficult for the Neanderthals as for us, involving the same unique (to humans) twisting of the baby's body during delivery.

In another CT study of the Tabun woman's pelvis, this time without a direct newborn comparison, the paleoanthropologists Tim Weaver and Jean-Jacques Hublin came to rather different conclusions, arguing that the Neanderthal birth process would not have been the same as ours. The modern birth canal is widest across its breadth higher up, but then alters downward to become widest front to back, which is why our babies generally change position as they descend. However, in their reconstruction Tim and Jean-Jacques found that the Tabun birth canal was wide across its breadth throughout, and thus the Neanderthal birth process would have been simpler than ours, without the need for additional rotations, and perhaps less dangerous. We *Homo sapiens* have narrower pelvises than either our Neanderthal cousins or our African predecessors, for reasons that are still unclear, but this evidence suggests that the change in our hips led to new evolutionary demands, probably requiring both biological changes in the process of delivery and social changes in the level of support needed for modern human mothers giving birth.

As we saw, teeth are a valuable resource in studies of our evolution, and because they are already highly mineralized, they preserve very well as fossils. Their size and shape are largely under genetic control (identical twins have similar teeth), and the form of the tooth crown has proved particularly useful in comparing fossil and recent humans. Distinctive patterns of tooth cusps and wrinkles characterize different populations today; a set of unworn teeth can be assigned to a region of the world with a fair degree of confidence using forensics. The anthropologist Christy Turner used this variation to propose an "Out of Asia" scenario for recent human evolution twenty years ago, based on the fact that the "average" dental morphology today can be found in the aboriginal peoples of southeast Asia. He argued that these populations were closest to the original modern human dental pattern, and that this indicated the location of the original source area for *H. sapiens*.

However, Christy's approach could not account for the undoubted similarity between recent Australian and African dental patterns, and I, together with my colleagues Tim Compton and Louise Humphrey, added fossil teeth from Europe to the mix, showing that an African origin for our present dental variation was still the most likely. That conclusion has been further strengthened by Christy's former students Joel Irish and Shara Bailey, who have added many other fossil teeth to their analyses. This kind of work has also been important in studies of earlier human evolution, for example, in showing that the Atapuerca Sima de los Huesos fossils are clearly related to later Neanderthals, and that the Skhul and Qafzeh early moderns from Israel have "African" traits in their teeth.

The famous British archaeological site of Boxgrove, near Chichester, has produced over four hundred beautifully made flint handaxes from levels also rich with the remains of interglacial mammals such as horse, red deer, elephant, and rhino. The fact that even the rhino bones showed extensive evidence of butchery led to a reevaluation of the capabilities of

Most of the four hundred handaxes from Boxgrove,
with the British Museum curator Claire Fisher.

hunter-gatherers 500,000 years ago, in terms of their primary access to
such resources. These people were not merely scavenging; they were apparently also highly capable hunters. They could secure the carcasses of large
mammals for the extraction of the maximum nutritional benefit in a landscape populated by dangerous competitors such as lions, wolves, and large
hyenas.

The importance of Boxgrove was heightened by the 1993 discovery of
a human shinbone attributed to *Homo heidelbergensis*, and two years
later, two lower incisor teeth from another individual were discovered.
Work with conventional light microscopes and scanning electron microscopes revealed a great deal of the evidence of animal bone butchery and
showed that the Boxgrove tibia had been gnawed at one end by a medium-
sized carnivore such as a wolf or a hyena. Microscopic studies also showed
that the front surfaces of the incisors were covered in a mass of scratches
and pits, suggesting that stone tools were being used as part of food processing, and the teeth were probably being marked accidentally during
such activities. Together with Mark Roberts and Simon Parfitt, the lead-
ers of the excavations, and the anthropologists Simon Hillson and Silvia
Bello, I was involved in further research on the incisors using a sophisticated imaging microscope, the Alicona.

These studies revealed other, perhaps less routine, activities. The teeth were certainly heavily worn on their crowns, suggesting that this was a middle-aged adult at death, but immediately apparent just below the crowns, much of the roots were coated in the sort of hard plaque that your dental hygienist is likely to remove during checks. This deposition indicates that the roots of these teeth must have been partly exposed above the gums during life, indicating receding gums or, more likely, that the front teeth were being strongly rocked back and forth, probably as they clenched something between them. For many years it has been argued from the strong and rounded wear on their front teeth that Neanderthals indulged in such behavior, and that food, fibrous materials, or skins were being softened or otherwise processed, with clenched teeth acting as a third hand, or a vice.

So it certainly looks like this activity has a much deeper antiquity in Europe, and that many of the cuts and scratches on the Boxgrove incisors were made unintentionally, when a flint tool cut through material being held in the mouth. But the Alicona revealed something else: there was also an unusual series of relatively fresh, deep, and semicircular scratches

The famous skeleton discovered in the Neander Valley, Germany, in 1856.

on the front surfaces of both incisors, evidently made near the time of death with much greater force than the other scratches, and in a completely different direction and action. The roots were also marked by heavy cuts, indicating that these too were made near the time of death. This suggests the possibility that these more violent actions were part of the butchery of this Boxgrove individual around (and we hope for his or her sake after) the time of death.

As well as carrying such scars of life and perhaps death, our teeth, as I already explained, contain important signals of our life history in their incremental lines, the dental equivalent of tree rings, which are laid down daily, rather than yearly. These lines have been studied microscopically through their surface expressions—such as the perikymata—but they can also be examined internally through broken surfaces or sections of teeth. I worked in a collaboration with anthropologists including Chris Dean, Meave Leakey, and Alan Walker to examine the growth lines in the molars of several fossil humans, including the Tabun Neanderthal from Israel, when a chip of enamel was briefly removed to apply electron spin resonance dating (see chapter 2) to it. We found that, unlike the pattern in tooth fragments of *Homo erectus*, this Neanderthal did overlap with the fastest developmental rates that we could find in modern molars. In 2007 another team, including Tanya Smith and Jean-Jacques Hublin, studied several teeth from a Neanderthal child from Scladina Cave in Belgium, which in terms of modern human dental development should have been nearly eleven years old when it died. However, their study determined that its actual age at death was only about eight, and the second molar was erupting significantly earlier in the Neanderthal than in modern children, thus signifying a shorter childhood and faster growth than ours.

It was unclear whether these different results regarding Neanderthal maturation were due to inaccuracies in the different methods, to variation between individuals, or perhaps even to evolutionary changes among Neanderthals in their growth patterns. What was most needed to resolve these questions were larger and wider-ranging samples from the fossils. However, as long as microscopic techniques were dependent on having naturally broken teeth or, even less likely, a museum curator willing to have his or her precious fossils sliced up, it seemed unlikely that such samples would materialize. And in terms of nondestructive techniques, only the very finest CT scans—microCT—could even begin to reveal the minute hidden details of incremental lines; so anthropologists have been very fortunate indeed to have yet another technology available to them: the synchrotron.

Many people have heard of the Large Hadron Collider, the world's largest and highest-energy particle accelerator, buried in a tunnel near Geneva, in Switzerland. This is a massive example of a synchrotron, a circular chamber that progressively accelerates atomic and subatomic particles such as electrons or protons, using electrical and magnetic forces. Not far away in Grenoble, France, is a smaller device that is occasionally diverted from the problems of high-energy physics to send its expensive electrons through precious fossils. The 52 kiloelectron-volt synchrotron X-ray beam has already revealed new species of beetles and ants from the time of the dinosaurs, entombed in opaque amber, and even tiny embryos of the dinosaurs themselves, within their mother's eggs. But the technology is now also being applied to hominin fossils such as the skull of *Sahelanthropus*, perhaps an ancestor from over 6 million years ago, and to more recent *erectus* and Neanderthal fossils. Resolution can be fourfold better than the best CT scanners, down to the width of a single cell, and researchers are now queuing up to submit their fossils to the magic of the Grenoble synchrotron.

In one of the first really significant uses of the synchrotron for modern human origins research, some of the team who announced that the Scladina Neanderthal had matured faster than we do were joined by the synchrotron researcher Paul Tafforeau, applying the technology to an early *Homo sapiens* child's jaw. This was from the Moroccan site Jebel Irhoud, and, as I previously described, one of the adult skulls from there was important in my realization that Africa could have been a key region for modern human origins. The Irhoud material is currently dated to about 160,000 years, and could be even older, but opinions differ about the classification of the specimens. In my view, overall, they still lie beyond the range of modern human anatomy and are farther away than specimens of a similar age from African sites like Omo Kibish and Herto.

Jean-Jacques Hublin, however, points to the fact that the child's jaw

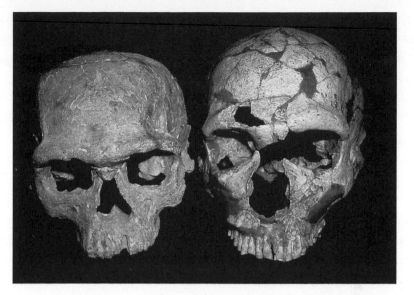

An early *Homo sapiens* skull from Jebel Irhoud (*left*) and
a Neanderthal skull from the cave of La Ferrassie, France.

has a chin, and there are some modern features in the face or braincase
of the two adult skulls from the site. But regardless of those debates, the
synchrotron study, which combined the counting of external growth lines
microscopically with the examination of hidden daily incremental lines
using the synchrotron, concluded that this child was about eight when it
died and was growing slowly, like a modern child. So even if, as I believe,
this was not a fully modern child anatomically, it was already growing its
teeth like one, with all the implications that this brings in terms of a prolonged childhood, deferred energetic demands, and an increased capacity
for learning.

My museum colleague Robert Kruszynski recently took the Devil's
Tower Neanderthal child's skull to Grenoble, and I joined with Tanya
Smith, Paul Tafforeau, Jean-Jacques Hublin, and other colleagues in what
we hope will be a definitive study of dental development patterns in Neanderthals and early modern humans. Data from many fossils were quanti-
fied and analyzed, and nine Neanderthals at different stages of maturity
were synchrotroned. These were then compared with similar analyses on
five early moderns and a large sample of recent people from different

regions. The results seem to finally establish that early moderns such as those from Skhul and Qafzeh were maturing dentally at the slow rate of recent humans, while the Neanderthals were growing somewhat faster, particularly in the case of the later erupting teeth. Thus, for example, the Neanderthal children from Engis, Scladina, and Le Moustier should have been about four, eleven, and sixteen years old respectively from modern development patterns, but the synchrotron showed that they were actually about three, eight, and twelve. Not only were the Neanderthals maturing faster, but the more rapid growth of their molars meant that these teeth had thinner enamel than ours. As we will see, such differences in life histories may have been significant in the distinct social structures and cultural development of Neanderthals and modern humans.

At an even finer level of our physiology than daily growth lines in our teeth, our bodies are built and maintained from the nutrition that we take in every day from food and drink. In that sense the old macrobiotic slogan "you are what you eat" is absolutely accurate. The many different chemicals in our foodstuffs are taken up into our bones and teeth and, if those are preserved as fossils, they can provide signals that can be interpreted as evidence of former diets. As mentioned in the previous chapter, atoms of substances such as carbon and nitrogen (two vital components of our bodies) come in the form of distinct isotopes, which have different atomic weights (because they contain different numbers of particles called neutrons). Although the essential properties of these slightly different atoms remain constant—for example, in the chemical compounds they can form—the compounds may behave somewhat differently when they are subjected to heat, say, where lighter isotopes may be preferentially evaporated. When compounds are taken up or pass through living systems, the rate at which this happens may vary between lighter and heavier isotopes. Carbon and nitrogen both have stable isotopes (that is, they do not undergo radioactive decay as the isotope carbon-14 does). These stable isotopes are found in collagen, a fibrous structural protein that makes up much of our body tissues and bones. Although bone and the collagen in it are constantly renewed as a living tissue, the turnover is quite slow, and stable isotope data from collagen will represent an average of diet over the last decade or more of an individual's life. Collagen is lost as a bone fossilizes, but enough can still be preserved in remains that are less than 100,000 years old for its constituent isotopes to be measured.

The relative abundance of the stable carbon isotopes C-13 and C-12 varies in different ecosystems, such as on land or sea, and also between different kinds of plants, so animals that feed from these different sources will pick up different ratios of the isotopes, according to their diets. In addition, the relative abundance of the stable nitrogen isotopes N-15 and N-14 increases by about 2 to 5 percent in each step up the food chain (for example, in moving from grass to the rabbit that eats it, and then to the human that eats the rabbit). So by simultaneously measuring carbon and nitrogen isotope ratios in the fossil bones of, say, a Neanderthal, it is possible to reconstruct something of that individual's diet during his or her life. The isotopes will not reflect everything, since the signals represent the main dietary protein sources rather than all foods consumed, and, as mentioned, the isotope uptake is also averaged over the last years of life. In addition, only broad categories such as the predominance of plant, herbivore, and carnivore protein derived from terrestrial or marine ecosystems can be distinguished. Care also has to be taken in analyses, as there is evidence that differences in climate such as temperature and aridity can affect the underlying stable isotope ratios, and thus the most reliable comparisons are on material relatively close in time and space.

Nevertheless, despite all these provisos, valuable insights into ancient diets have been obtained from the bones of both Neanderthals and early modern humans by researchers like Michael Richards and Hervé Bocherens. Over a dozen Neanderthals and even more Cro-Magnons have been analyzed, and clear patterns have emerged that confirm our view that the Neanderthals were heavily dependent on meat from large game such as reindeer, mammoth, bison, and horse. They were at the top of their food chains, and their isotope signatures place them with wolves and lions as the dominant predators in their landscapes. However, the fossils analyzed come from regions like France, Germany, and Croatia, and do not yet cover the whole Neanderthal range—unfortunately, warmer regions like Gibraltar and the Middle East have poorer collagen preservation. And we know from archaeological data that farther south, in the coastal regions of

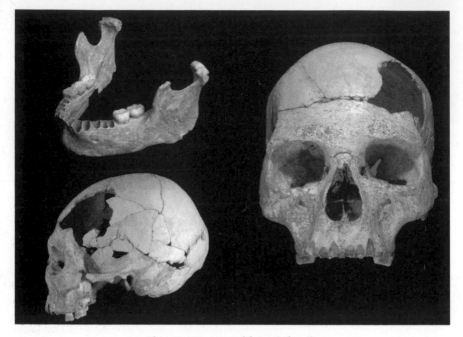

The 40,000-year-old Oase fossils.

Portugal, Spain, Gibraltar, and Italy, the Neanderthals were supplementing their big game with marine resources such as shellfish, seals, and, at least occasionally, dolphins that had probably been stranded. Although the isotope signal would have been obscured if meat was the dominant source of protein, where conditions and the seasons allowed, plant resources were also important to them, as burned nuts and seeds in the cave deposits show.

But analyses of the Cro-Magnons, including the rather primitive 40,000-year-old Oase early moderns from Romania (see the next chapter), present a different dietary picture to that of the Neanderthals, even when comparisons are limited to the same regions and climates. High levels of C-13 in some of the samples near coastlines suggest a menu that included significant quantities of marine fish or other seafoods, while those that lived farther inland may have had even more diverse diets, since unusually high levels of N-15 in their bones suggest that fish, waterfowl, and other freshwater resources provided important sources of food. And the fact that this was already the case in the oldest modern human sampled, from Oase, becomes even more significant when we look at another modern fossil of similar age over 6,000 kilometers away in China: the adult skeleton from Tianyuan Cave, recently excavated in the Zhoukoudian complex of sites. (This skeleton is further discussed in the next chapter.) Carbon and nitrogen isotopes in collagen extracted from the bones indicated a diet high in animal protein, but the very high nitrogen isotope ratio also suggested the consumption of freshwater fish. Sulphur isotopes were then measured in terrestrial and freshwater animal remains from ancient and recent archaeological sites in the Zhoukoudian area to give a baseline comparison for this additional dietary indicator, and when the sulphur values were analyzed for the Tianyuan skeleton they confirmed a substantial portion of this individual's diet must have been made up of freshwater fish.

So with the very earliest evidence we have of modern humans in their dispersal from western Asia into Europe and into the Far East, the abilities of *Homo sapiens* were already apparent in the extraction of more nutrition from their environments than the Neanderthals could achieve, and this was surely one of the keys to our survival and eventual success in the challenging environments of the north. The ratio of two other stable isotopes, of strontium (Sr-87 and Sr-86), may also help us track some of the migrations of those early humans, as the relative proportions of these isotopes vary in the different ground rocks through which water flows. When animals, including humans, ingest these isotopes in their diet or drinking water, they are taken up into the bones and teeth in the same way that calcium is, and thus provide a marker of where an animal had been living at the time or earlier in life. Tooth enamel will preserve an indicator from childhood, while bones, with their turnover, will record a signal from the last few years of life, so by comparing the isotope ratios in a fossil skeleton with those in the rocks of the surrounding landscapes, it is now theoretically possible to "locate" a Cro-Magnon to, say, its local limestone area or perhaps to granite hills many kilometers away, where that individual may have lived as a child. And if that Cro-Magnon was eating reindeer, it would be possible to map the migrations of those reindeer herds from their remains in the same archaeological site. This technique is now becoming feasible for precious fossils because, with improvements in technology and measurement, laser ablation can produce isotope results from even minute fragments of enamel.

Teeth and bones contain many clues to the life histories and activities of long-dead people. Having discussed how new technologies are helping us to date and investigate important relics, let us now turn back to the fossil record itself, to pick up the developing story of the origin of our species.

4

Finding the Way Forward

IN CHAPTER 1 I SHOWED HOW RAO WENT FROM BEING A VIEW that no one held in 1970 to the dominant model for modern human origins in an astonishing period of less than thirty years, and I laid out the standard rapid African origin, expansion, and replacement view that people like me began to develop and hold from around 1984. Now I'd like to look at how some elements of the early part of that scenario are taking on a different significance, challenging the orthodoxy of Out of Africa 1, with implications for our own ultimate origins.

In 1991 surprising new finds started to be made in western Asia, at Dmanisi in Georgia. An ancient village on a hill was being excavated by medieval archaeologists when they discovered the remains of a rhinoceros jaw in the cellar of one of the buildings. This was just about explicable if a traveler had brought or traded a specimen back from Africa or Asia, but expert opinion determined that it was actually a fossil rhinoceros, perhaps a million years old, and that was a lot more difficult to explain! It turned out that the village had, by chance, been built on a much more ancient fossiliferous deposit, and once the medieval archaeologists and the paleontologists had agreed how the site could be dug to the satisfaction of both parties, new excavations got under way. Pleistocene fauna was found, then a human lower jaw and primitive stone tools. Georgian workers and their foreign collaborators argued that the site was potentially about 1.8 million years old, but everyone else was cautious, since such an age challenged prevailing views, and when we got our first sight of the fossil at a conference in Frankfurt in 1992, most of us thought that the jaw seemed too evolved for such a great age. However, two Spanish researchers, Antonio Rosas and José María Bermúdez de Castro, reported that the jawbone resembled both early *erectus* specimens from East Africa and later *erectus* material from China. Further excavations and research amply confirmed the original claims, placing the date at about 1.75 million years, and producing five small-brained human skulls, three more jawbones, many other parts of the skeleton, and a quantity of very basic stone tools, often made from local volcanic rocks. These finds were, and still are, challenging, as it used to be thought that the first move out of Africa was enabled by changes in behavior, increasing brain size, or better tools—and none of those developments seem to be evidenced at Dmanisi. Some of the animals had probably dispersed from Africa, including two large saber-tooth cat species. These specialized animals lacked the teeth to strip a carcass clean of its meat or break the thicker bones of their prey, so they potentially provided scavenging opportunities for the early humans from what they left behind. But wider comparisons of the animal species suggest that the Dmanisi assemblages most closely resembled those of the grasslands and woodlands of southern Europe at that time, supporting the idea that these early non-Africans had already adapted to new environments.

The second find I am going to discuss presents even greater challenges to conventional thinking on human evolution, so much so that one expert implied it is more like the Piltdown hoax than a genuine fossil relic! It used to be thought that only one species of early human lived in southeast Asia before modern humans arrived there: the ancient species *Homo erectus*, best known from the island of Java, as we saw. *H. erectus* could have reached Java from southern Asia at times of lower sea level, when the islands formed part of a larger ancient landmass that scientists have called Sunda (from an Indonesian word for western Java). Without boats, *erectus* could get no farther. Thus it was generally believed that Java/Sunda represented the farthest limit of human colonization in the region until modern humans arrived, perhaps 50,000 or 60,000 years ago, who were able to use boats to disperse even farther, toward Australia and New Guinea. But in 2004 remarkable evidence was published by the

Australian archaeologist Mike Morwood and his team of a new human-like species from the island of Flores, about five hundred kilometers east of Java. The remains included much of the skeleton of an adult estimated to be only about a meter tall, with a brain volume of about four hundred cubic centimeters (about the same as that of a chimpanzee). This find and other more fragmentary specimens were discovered in Liang Bua Cave on Flores, associated with stone tools and the remains of a pygmy form of an extinct elephant-like creature called *Stegodon*. The skeleton was dated to only about 18,000 years ago and was assigned to a new species called *Homo floresiensis* ("Flores Man"), but it soon became better known by its nickname "the Hobbit."

I was privileged to be chosen as the main commentator at the press conference in London that launched the startling new find to the world, but it was unexpected for so many reasons. It lay five hundred kilometers of islands and water beyond the accepted range of ancient humans, implying that the Hobbit's ancestors must have had boats to get there; it seemingly had a "human" face and teeth, and walked upright, yet it had an ape-sized brain; despite its small brain it was associated with stone tools and possible evidence of hunting and fire. If it was a distinct species, where did it come from, how did it survive long after other forms like the Neanderthals, and what happened to it after 18,000 years ago? There was immediate and fierce controversy about the nature of the finds, and whether they had been correctly interpreted. Some scientists insisted that they were wrongly dated and might represent small-bodied modern humans; others argued that the unusual features were signs of disease, perhaps due to medical abnormalities such as microcephaly, cretinism, or a condition called Laron syndrome. The situation was exacerbated when the late Teuku Jacob, an esteemed but retired Indonesian paleoanthropologist (who was a Hobbit skeptic, and not part of the original research team), "borrowed" the finds in order to conduct his own studies. When they were eventually returned, in the face of vigorous protests, some of the bones had been seriously damaged, apparently through hurried and botched attempts to replicate them.

The authors of the original studies suggested that *H. floresiensis* might be a descendant of *H. erectus* that had arrived earlier on Flores, perhaps using boats. This idea was partly inspired by the existence of stone tools on the island that were at least 800,000 years old. They argued that the species then evolved a very small size under isolated conditions—a phenomenon known to occur in island populations of other medium-large-sized mammals and called *island dwarfing*. If the ancestors of *H. floresiensis* really made watercraft to reach the island, this would be surprising, because such behavior is usually considered to be exclusive to our species. (Even the Neanderthals were seemingly unable to cross the English Channel from France about 120,000 years ago, or reach the islands of the Mediterranean, with the possible exception of Crete.) However, the alternative of accidental rafting on mats of vegetation must also be considered; the Asian tsunami of 2004 dispersed people on rafts of vegetation for over 150 kilometers, reminding us that this is a feasible process in a geologically active area like Indonesia. Moreover, studies of the prevailing ocean currents suggest that the ultimate origin of the Hobbit may not have been from Java to the west, but from Sulawesi to the north. Mike Morwood's follow-up work in Sulawesi lends support to this idea, since stone tools that could be at least 1 million years old have also been found there, although sadly so far without accompanying fossils.

Further studies provided more detailed information on the limb bones of *H. floresiensis*, both from the original skeleton and from other individuals, all of them very small, and some dating as far back as 95,000 years. A second jawbone, similar in its primitive and distinctive features (for example, its lack of chin, thick body, and divergent tooth rows) to the one associated with the original skeleton, was described by the anthropologist Peter Brown. Puzzlingly for a supposed human species, the body proportions, wrist bones, hip bones, and shape and robustness of the arms and legs of *H. floresiensis* are in some ways more similar to fossils of prehuman African species like *Australopithecus afarensis* (the most famous example of which is "Lucy") and the newly discovered *Australopithecus sediba* (from South Africa) than to later humans. In addition, unusual features of the shoulder joint have been reported, as well as the fact that *H. floresiensis* seems to have had large flat feet! While these peculiar features have fueled speculation that the remains are abnormal, other workers have argued that they show evidence for an unusual evolutionary trajectory, in island isolation. In the case of the wrist, the shape of these bones in two different adult individuals resembles much more

closely those of apes and *afarensis* than of recent humans such as Neanderthals and us—and these are bones whose shapes are already mapped
out before birth. The likelihood that a pathology in two different individuals could independently convert wrist bones back to a similar primitive-
looking condition seems remote indeed.

Further uncertainty surrounds the evidence of humanlike behavior
excavated from Liang Bua. Some of the stone tools are delicately shaped,
and there is possible evidence of burning (although perhaps naturally
produced) and of predation on young *Stegodon*. I'm still not convinced
that *H. floresiensis*, with its ape-sized brain, was capable of such behaviors,
and in my view we need further evidence and analyses to exclude the possibility that early modern humans were also using caves on Flores before
18,000 years ago and could be responsible for some of the later archaeological evidence left behind—although this is unlikely to be so in the
deepest levels of the cave. If *H. floresiensis* is indeed genuine and distinct,
rather than abnormal (and I think the evidence is growing strongly in its
favor), there are the intriguing questions not only of where it came from
(Java to the west or, as Morwood now believes, Sulawesi to the north?) and
how it got there, but of what happened to it, and whether our species
encountered these diminutive relatives. Perhaps volcanic eruptions or climatic change around 17,000 years ago affected its habitat, or perhaps modern humans killed it off directly, or by consuming the resources on which
it lived. In which case, an even stranger encounter than the one between
the Neanderthals and modern humans was played out at an even later
date, on the opposite edge of the inhabited world. But against that possibility is the evidence that there may instead have been a period of several
thousand years with no one on Flores, following the extinction of the Hobbit, before modern humans finally arrived after 12,000 years ago.

The Hobbit remains a perplexing find for all of us, whatever our evolutionary views—witness my own problems in coming to terms with the
possibility that its chimp-sized brain could be associated with "human"
behavioral complexity. But it has proved most difficult for some scientists
of multiregional persuasion who are wedded to the idea that there can only
have been one human species—*Homo sapiens*—in existence over the last 2
million years or so. Rather than contemplate a painful divorce from cher-
ished beliefs, they have preferred to argue that the Hobbit represents the

"village idiot" of a modern human community—or even more remarkably, that it's a recently buried oddity, as shown by the presence of dental fillings (there is no evidence that the Hobbit was ever seen by a dentist!). In the short term these researchers have raised their profiles by courting controversy, but in the longer term, I think they have damaged their own, and paleoanthropology's, reputation.

As we saw, under Out of Africa 1, most experts consider that *H. erectus* was the first humanlike creature to emerge from the ancestral African homeland, nearly 2 million years ago. But for some researchers, the Flores material raises the possibility that more primitive, perhaps even prehuman, forms had previously spread from Africa across southern Asia, where the remoteness of Flores allowed them to survive and evolve along their own peculiar path, in isolation. And the evidence from Dmanisi is now being added to this rethink, since the lack of very ancient fossil human evidence from Asia, apart from Dmanisi, is considered by archaeologists like Robin Dennell and Wil Roebroeks to reflect a lack of preservation and discovery, rather than a real absence. Combining the primitiveness of the Dmanisi specimens and tools with a similar view of the Liang Bua finds, it is argued that there was a widespread phase of human evolution in Eurasia about 2 million years ago, which is now only represented by the isolated Dmanisi and Hobbit fossils. This alternative scenario has a small-brained and small-bodied pre-*erectus* species, perhaps comparable to *Homo habilis* or even a late australopithecine, dispersing from Africa with primitive tools over 2 million years ago, reaching the Far East and, eventually, Flores. In Asia, this ancestral species also gave rise to the Dmanisi people and *Homo erectus*, while Dmanisi-like people reentered Africa about 1.8 million years ago and evolved into later populations there—including, eventually, *Homo sapiens*. So the orthodoxy of Out of Africa 1 is being challenged because of new evidence, and new interpretations of old evidence. And the same process of reevaluation is happening with Out of Africa 2, as we shall see next.

Views on the origin of our species have gone through many formulations and reformulations since Darwin laid out his expectations of what the evidence would provide, but an African origin for *Homo sapiens* is now the mainstream view. I explained earlier how finds like those from Dmanisi and Flores have threatened the scenario known as Out of Africa

1, and we will now look at new evidence from Europe, Africa, and Asia that is changing ideas about the more recent parts of our evolutionary story. These finds include the remarkable 160,000-year-old Herto skulls from Ethiopia (some of the oldest and most massive individuals of our species ever discovered), the 40,000-year-old fossils found by cavers deep in an underground chamber in Romania that may show hybridization between modern humans and Neanderthals, and the oldest *sapiens* fossils from China, whose feet hold clues to a vital modern innovation. In the last few years we have also learned a lot about our Neanderthal cousins: where they came from, how they behaved, how their bodies worked, and even (as I explain later) how their whole genetic code compares with ours. But now I'm going to highlight some of the most interesting fossil finds of Neanderthals from the last twenty years, before going on to discuss the evidence of other peoples who may have been closer to our evolutionary origins.

Asturias, just south of the Bay of Biscay, is one of the less fashionable provinces of Spain. But like much of the Iberian peninsula, right down to its southern tips in Portugal and Gibraltar, Asturias was a favored territory for the Neanderthals. In 1994 some explorers were probing the depths of the large and still partly unexplored El Sidrón network of caves, hidden among densely forested hills, when they discovered two human jawbones lying in the cave sediments. Because partisans were known to have hidden in the cave during the Spanish Civil War, the police were notified in case the remains were recent, and over a hundred further bones were soon discovered. Forensic investigations over several years showed that the bones were fossilized, not recent, and were in fact those of Neanderthals who had died over 40,000 years ago. The area where the bones were found has been named Galería del Osario—"tunnel of the bones"—and about 1,500 bone fragments from some twelve Neanderthals have now been unearthed there. At first glance it seems like an extended family was represented, since there are adults, teenagers, and children, but this was no happy domestic scene, at least not in the fate that seems to have befallen them. Their bones and teeth suggest they were reasonably healthy, although there are signs of growth disturbances during early and late childhood in the teeth.

However, the state of the fossilized bones shows that these individuals may have died violent deaths: they displayed many cut marks, especially

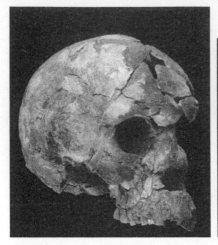 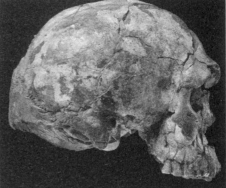

Oblique view of the most complete 160,000-year-old Herto *Homo sapiens* skull from Ethiopia.

Side view of the Herto 1 skull.

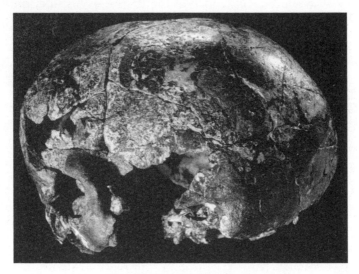

The child's skull from Herto.

one jawbone and the children's skulls, and they may have been pounded and smashed with great force, using stone tools or rocks, apparently to extract the nutritious brain and marrow. So this seems to be evidence, and by no means the first, of cannibalism among Neanderthals. Other examples are known from places like Croatia (Vindija) and France

(Marillac and Moula-Guercy), and seem to reinforce stereotypes of the Neanderthals as savage subhumans.

Yet cannibalism seems to have been a regular enough part of human behavior over the last million years or so for it to be represented in many fossil assemblages, and so it might almost be considered as "normal" for early humans, however distasteful (in every sense!) we might find it today. It appears to be present in the *Homo antecessor* ("Pioneer Man") remains at Atapuerca about 800,000 years ago, which are cut-marked and smashed like the Sidrón fossils, and which lie alongside butchered animal bones. It may also have been present in *Homo heidelbergensis*—at Bodo in Ethiopia about 600,000 years ago (although here it's mainly evident from cut marks on a skull suggesting the eyeballs had been removed), and at Boxgrove in Sussex, where I explained that two isolated front teeth seem to have been forcefully wrenched from their jawbone (now lost). Its antiquity may even stretch back to the very first humans. The cheekbone of a 2-million-year-old fossil skull from Sterkfontein (South Africa), often assigned to the very early human species *Homo habilis* ("Handy Man"), shows signs of having been cut during the slicing apart of the jawbone from the skull, and this too may have been for consumption. And we should not let our own species off the hook either, since 80,000-year-old bones from Klasies River Mouth Caves in South Africa, and 14,700-year-old bones that I helped to excavate from Gough's Cave in Somerset, also show the telltale signs of butchery. Unfortunately, there is also enough sound evidence beyond the exaggerations in traveler's tales to indicate that butchery and consumption of human flesh have occurred in the very recent human past.

Of course, we have to put these gruesome reconstructions of past behavior into context against the possibilities that disarticulation of bodies might also have occurred as a part of funerary rites, or that cannibalism was part of rituals to honor the dead, or was forced on human groups faced with disaster or starvation, as recent history demonstrates. We also have enough evidence of the care of individuals during life, and after their death, to show the other side of the story, as I will discuss in more detail in chapter 6. For example, a child in the Atapuerca early Neanderthal material from the Sima de los Huesos site in northern Spain had severe head deformities that certainly affected its appearance, and probably also its behavior and speech, yet it had survived well beyond the infant stage. At times the Neanderthals buried their dead in caves, from newborn infants through to elderly men and women. In at least some cases, grave goods such as animal remains and special rocks or tools seem to have been placed with the bodies, as tributes or perhaps even in anticipation of an afterlife.

Notwithstanding those caveats, which show the compassionate face of our predecessors, I think that early humans were probably as capable of love and hate, and tenderness and violence, as we are, and even chimpanzee bands have been observed in violent and often fatal territorial "battles" with other troops. Such behavior is almost certainly part of our evolutionary history too. Many years ago, in an obscure book, *The Dawn Warriors*, partly inspired by Robert Ardrey's *African Genesis* (itself very much a tale of humans, red in tooth and claw), the biologist Robert Bigelow argued that warfare went back to the beginnings of humanity and had shaped our evolution. Coping with conflict from other human groups encouraged individual intelligence and cunning, and group cooperation and cohesion, and thus fueled social evolution, language, and the growth of the brain. This is something I will come back to in chapter 6, but we have not yet finished with the unfortunate Neanderthals of El Sidrón.

After they had died and apparently been eaten, their defleshed remains must have lain on the floor near the cave entrance, along with other food debris and Middle Paleolithic (Middle Old Stone Age) stone tools, perhaps including those used to butcher them. There the bones would most likely have been trampled, eroded, or scavenged by other animals. But then, serendipitously, a massive muddy collapse of cave sediments dropped them some twenty meters deeper into the cave system and dramatically increased their chances of long-term preservation, in the cooler location where they were ultimately discovered. This also greatly improved the potential for DNA preservation; the Sidrón Neanderthals are now one of the most important contributors to the Neanderthal Genome Project, as I will discuss in chapter 7, along with the main DNA donors, other probably cannibalized Neanderthal from Vindija Cave in Croatia. There are other finds, however, that give a different and more positive perspective on the Neanderthals than this image of cannibalism, and one of the most significant of these was made about thirty years ago, in the French site Saint-Césaire.

The discovery of a partial skeleton in the collapsed rock shelter of Saint-Césaire remains one of the most important of all Neanderthal finds—not just because it was fairly complete by the usual standards, and not because it seemed to be a burial—there were quite a few of those already known for the Neanderthals. Its importance lay in its archaeological associations: stone tools belonging to the Châtelperronian industry. This enigmatic industry from southwestern France seemed to represent a transition from the local Middle Paleolithic (Mousterian) to the Upper Paleolithic Gravettian. The local Mousterian and the Châtelperronian had many types of stone tools in common, but the way they were made had seemingly switched from typical Neanderthal flaking to the systematic striking of thin flakes—blades—something characteristic of the Upper Paleolithic, and thus of Cro-Magnons like the Gravettians. Unfortunately, there were no human fossils reliably associated with the Châtelperronian, which is why its real significance remained a puzzle. Many archaeologists and anthropologists in the 1960s and 1970s expected that the manufacturers of the Châtelperronian would, when eventually discovered, turn out to be evolutionary intermediates between the Neanderthals and the Cro-Magnons, hence finally proving the Neanderthal Phase and Multiregional models of modern human origins. The archaeologist Richard Klein and I shared a different view. We thought that Neanderthals were probably capable of making Upper Paleolithic–style tools and, considering the local ancestry of the Châtelperronian, that the manufacturers were likely to have been Neanderthals and not transitional forms.

So when I heard of a brief French report in 1980 that a human skeleton had finally been recovered with Châtelperronian artifacts, I realized that this could be a crunch discovery that might completely invalidate my doctoral conclusions that there was unlikely to have been evolutionary continuity between Neanderthals and Cro-Magnons in Europe. I waited for news of the nature of the skeleton with bated breath, and I must admit to considerable relief when it was identified as a rather typical Neanderthal! However, to my chagrin, many researchers seemed reluctant to abandon ideas of continuity. Some, like Milford Wolpoff, argued that

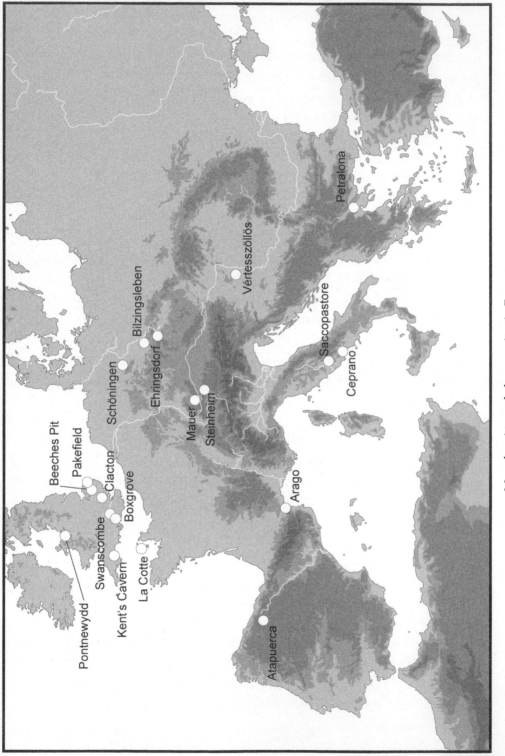

Map showing early human sites in Europe.

the Neanderthal features of the Saint-Césaire skeleton had been overemphasized and that it was, in fact, "transitional," while others, like the archaeologist Randy White, suggested that (in line with Loring Brace's Neanderthal Phase model) cultural change had probably preceded, and driven, morphological changes toward modern humans. Thus this Neanderthal had not yet undergone the evolutionary transition that must have followed.

For a few years the significance of Saint-Césaire was a hot topic of debate, but gradually, as its Neanderthal nature was generally accepted, it became an important piece of evidence supporting the Replacement model, at least in western Europe. This was because the Châtelperronian had been dated by radiocarbon to about 35,000 years—the same age as the other early Upper Paleolithic industry, the Aurignacian—which seemed to be associated with modern-looking Cro-Magnons. Hence several of us, including the archaeologists Richard Klein and Paul Mellars and the anthropologist Bernard Vandermeersch (who had described the new find), favored a model with two parallel but distinct strands within the early Upper Paleolithic of western Europe. One, the Châtelperronian, was a local Neanderthal development. The other, the Aurignacian, was the product of modern humans (Cro-Magnons), who had brought it with them when they entered western Europe, an event now dated to about 40,000 years ago. But now I'd like to discuss a remarkable site in eastern Europe where modern human fossils have turned up alongside the remains of thousands of cave bears, which suggests a previously unknown and perhaps even earlier arrival of modern humans in Europe.

The Danube is, at some 2,900 kilometers, Europe's second-longest river (after the Volga). Originating in Germany, it flows eastward until it reaches its delta, which spreads between Romania and Ukraine on the Black Sea. It had been a hugely important waterway in historic times, and it must also have provided a route through the landscape for early humans, whether they trekked along its banks or (later on) used rafts or boats to navigate it. A number of key sites in the Neanderthal and early modern story lie close to the Danube, and one of the newest and most fascinating of these is Peştera cu Oase ("Cave with Bones"). Oase was discovered by cavers in 2002, and its location is still a closely guarded secret, although it lies in the Carpathian Mountains of western Romania, whose rivers drain into the Danube.

One of those rivers is the Ponor, which runs for about 750 meters underground, and above it are networks of caves through which it formerly flowed. These cannot be reached through their original entrances, which have long since been blocked by sediments and collapses. Instead, a lower entrance has been opened up by speleologists, and, for those brave and capable enough, a trip up and down long shafts, with a scuba dive in the dark through a sixteen-meter siphon of chilling water thrown in for good measure, eventually leads to a cave floor littered with an astonishing collection of thousands of fossil bones. Among these are circular hibernation nests of cave bears, the former winter inhabitants of this part of the cave, represented by over a hundred of their skulls alone. Other occasional residents such as cave lions and wolves are also recorded, but it was a chance discovery in a neighboring chamber in 2002 that showed ancient humans had also been in the vicinity. This was a human lower jaw, containing only the back teeth, with the presence of erupted wisdom teeth showing that the individual was an adult, and from its overall size probably a young man.

The following year, about fifteen meters downslope, the face and skull bones of a younger individual were also found. These isolated human finds seem to have been washed to where they were discovered, since there is no evidence for human occupation or human interference with the bones in this part of the cave system. Careful mapping, excavation, and dating (including direct radiocarbon determinations on the human fossils) suggest the following sequence of events for the cave system. Cave bears regularly entered the deep cave to hibernate until about 46,000 years ago, with many dying during hibernation. About 46,000 years ago, there was a major collapse, which changed the nature of the cave by opening up a closer entrance. Carnivores such as wolves and the occasional lion now denned there, bringing back the remains of prey such as deer and ibex. About 42,000 years ago, the Oase humans were in the cave, either using it for shelter or perhaps carried there by carnivores, and their remains were then swept to where they were discovered, in one of the periodic floodings that engulfed the cave system.

An international team of researchers has been studying the Oase finds, and in particular the human jawbone and the separate skull. The jawbone is strongly built but undeniably modern in its well-developed chin. And yet its molar teeth are large, with quite complex cusps, and it has some interesting features toward the back. The ascending ramus of the jaw is extremely broad, and on its inner side there are yet more intriguing traits. On the inside of the back of this jawbone on each side is the *mandibular foramen*, a hole through which the mandibular nerve travels to the lower teeth. In most living humans and in almost all fossil ones, the foramen is open and V-shaped, as it is on one side of the Oase jaw. But on the other side there is a bridge of bone across the foramen, known as the *horizontal-oval* (*H-O*) type. This H-O foramen is present in about half of all Neanderthal fossils, but it is generally rare in early modern fossils and usually only occurs at a level of a few percent in living humans. Accordingly, when found in a smattering of European Cro-Magnon fossils, it has been seen as a marker of possible Neanderthal ancestry, which it certainly could be, given its rarity in preceding African fossils and those from Skhul and Qafzeh. How much its presence is inherited, and how much is associated with having particularly strong jaw ligaments (which are attached to that area of the jaw), is still unclear, and the picture is further confused by the fact that some of the highest frequencies of it in recent populations occur far away from Europe and Neanderthal influence, in places like Easter Island.

The Oase skull is not Neanderthal-like, but it's strangely unlike later Europeans too. It's from an adolescent, of uncertain sex, with the third molars still unerupted but enormous, and with even more complex cusps than in the separate mandible. The forehead retreats a little, but the lack of brows and suprainiac fossa and the flat face and nose shape are particularly non-Neanderthal. However, the back teeth are bigger than those of any modern humans I have seen from Eurasia, living or fossil. In support of his favored Assimilation model, my friend Erik Trinkaus argued that features like the H-O foramen, the relatively flat forehead, and the large molars could all be a sign of a mixed modern–Neanderthal heritage. And indeed, at a date of more than 40,000 years, the Oase humans seem to have been in the vanguard of modern human penetration of Europe, with the maximum potential for encountering Neanderthals. On the other hand, to me, the teeth look unlike those of Neanderthals as well, so where did these enigmatic early Europeans come from?

At this point we should note that the Oase human fossils have no associated artifacts; even if they were living elsewhere in the cave, only their actual bones were recovered, following redeposition from an unknown location. But given their unusual features and their antiquity— perhaps beyond the age of other European moderns and definite Auri- gnacian tools—I think they could even have been makers of the mysterious Bohunician industry. This was named after the Czech cave Bohunice, and like the Châtelperronian, its characteristic tools show a mixture of both Middle and Upper Paleolithic elements. The method of manufacture is often the "prepared core" so typical of Middle Paleolithic artifacts in Europe, western Asia, and Africa; in that sense the Bohunician maintains an older tradition associated with both Neanderthals and the earliest modern humans. But there are also many of the blades, end scrapers, and burins (engraving tools) that characterize the European Upper Paleo- lithic and African Later Stone Age made by modern humans. However, the Bohunician has not, so far, shown any evidence of sophisticated bone or ivory artifacts, or beads.

So, as with the puzzle of the Châtelperronian, was the Bohunician made by Neanderthals or moderns, or someone in between? It has not yet yielded any diagnostic human remains, but there are interesting clues from its dating and possible origins. The Bohunician was dated by radio- carbon and luminescence methods to about 45,000 years—slightly older than the Oase humans—and certain aspects of the tools and the way they are made link with similarly aged industries in Turkey and the Middle East, as we shall see. Could this signal a previously unrecognized dispersal of modern humans into Europe before the Aurignacian? For me, clues to the origin of the Bohunician and the Oase people are not likely to be found at an older date in Europe—both seem not to have local antecedents and therefore look intrusive. And what was the ultimate fate of the Oase folk? Did they give rise to the succeeding Cro-Magnons of Europe? Or were their pioneering colonization of Europe and their possibly unpredictable encounters with the Neanderthals ultimately in vain, terminated by cli- matic and environmental catastrophes that soon followed?

As mentioned in chapter 2, around 39,000 years ago, there was a mas- sive eruption in the volcanic area of southern Italy known as the Campi Flegrei (the "burning fields"), near the Bay of Naples. The region is still active today and includes the Solfatara crater, the mythological home of the Roman god of fire, Vulcan. The eruption was perhaps second in extent to that of Toba in Sumatra during the last million years, and it produced the massive ash deposits known as the Campanian Ignimbrite (CI), which extend for some eighty kilometers. It also produced finer sulphur-rich deposits that spread even farther over an area of some 5 million square kilometers of the Mediterranean and western Eurasia, and it might also have produced a brief *volcanic winter*. Such a period of global cooling can be caused when increased atmospheric dust and droplets of sulphuric acid reflect sunlight back into space, and finer sulphide compounds may have a longer-term effect if they reach and stay in the upper atmosphere.

This episode was closely followed by a *Heinrich event*, a phenomenon first described by the geologist Hartmut Heinrich. During these brief but severe cold events, armadas of icebergs broke off the northern ice caps into the North Atlantic, for still unknown reasons. As they flowed south they melted, chilling the ocean and surrounding lands, and shedding ice-rafted debris to the ocean floor, one of the characteristic signals of these events in deep sea cores. "Heinrich 4" occurred about 38,000 to 39,000 years ago, and it severely chilled Europe, its influence showing in cores as far east as the lakes of Italy and Greece and the eastern Mediterranean.

Some researchers have even argued that this unusual combination of a volcanic winter, without proper summers for a number of years, closely followed by the chill of a Heinrich event drove many of the changes we can detect in the archaeological record of Europe around this time by either suddenly crashing (*bottlenecking*) Neanderthal human populations or forcing them to move, interact, and adapt in entirely new ways. In one view this opened the door for modern humans to colonize Europe, and in another interpretation it catalyzed dramatic changes in Neanderthal behavior and biology, converting the survivors into modern humans! I don't accept either explanation, because we know from Oase Cave in Romania and Kent's Cavern in England that modern humans were already in Europe before this time, and we know from sites in and beyond Europe that behaviors were already changing significantly. But there seems little doubt that these events would have profoundly affected both the Neanderthal and early modern human inhabitants of Europe. If the pioneering moderns were only in Europe in small numbers, while the Neanderthals remained in occupation of southern France and the southerly peninsulas of Iberia, Italy, and Greece, it seems likely that both *Homo sapiens* and their Neanderthal cousins would have suffered attrition. So perhaps the Oase people died out, just as many Neanderthals did at this time (see chapter 2).

Where did these newcomers come from, before they reached Romania? One clue lies far across Turkey, on its rugged coast, just fifteen kilometers away from the border with Syria. Üçağizli ("Three Mouths") Cave was discovered and first excavated in the 1980s, and now lies about eighteen meters above the Mediterranean. The sea floor slopes steeply away about five kilometers out, so even with the dramatically lowered seas of the last Ice Age, Üçağizli would never have been far from the coast. The cave deposits cover over 10,000 years, beginning about 44,000 years ago, and they contain many thousands of tools. The earliest of these resemble an industry in neighboring Middle Eastern countries called the Emiran, as well as the Bohunician of Europe, while subsequent tools from around 36,000 years represent something called the Ahmarian, about which more later in this chapter. But the site has many more features alongside the tools, including evidence of both short-term (small fire pits of charcoal) and long-term (huge heaps of ash) human occupation. One level even contains an impressive curved row of limestone blocks, perhaps a low wall, of unknown function.

An international team, including the archaeologists Steven Kuhn and Mary Stiner, has been excavating the cave since 1997 and has found not only large numbers of flint tools but also numerous bone points—perhaps piercers. Even more extraordinarily, they recovered many hundreds of ornaments made from shells, which must have been strung as beads or pendants. These were probably parts of necklaces or bracelets, and, in one instance, the talon of a vulture had also been utilized. While large shells in the site show signs of being smashed for consumption, the small ornamental ones are usually whole, apart from being pierced, and seem to have been collected from lake shores, rivers, or beaches expressly for the purpose of jewelry. From discarded failed attempts, the shells were probably worked at the site with pointed tools, and the holes were positioned very consistently, while some show signs of rubbing, where they were probably strung on natural fibers. In the case of a genus called *Dentalium* (tusk shells), these seem to have been collected as fossils from a geologic deposit about fifteen kilometers away and were snapped at intervals to create tube beads. Interestingly, the oldest layers in the cave show a dominance of lustrous *Nassarius* (tick shells), exactly the kind used as jewelry by much earlier modern humans in Africa and Israel (see chapter 5), suggesting a symbolic tradition running back 50,000 years earlier.

As we do today, the Üçağizli people probably used their appearance (including their ornaments) as conscious or unconscious symbols of group identity, marital status, and their roles in society. Equally, the use of such symbols implies that the meaning of what they were signaling would be recognized within their communities, and perhaps also when other human groups were encountered (see chapter 5). The cave obviously acted as a bead factory, but we don't know how significant this was socially to the people of the time. Perhaps it was just a convenient shelter or camp near the Mediterranean, a place where they could carry out the work, or maybe it fortuitously preserved the evidence of what were actually widespread activities better than more exposed locations. But based on the ornaments of modern hunter-gatherers, and what we find in the later Paleolithic sites of Europe, we can guess that these shell beads were only part of the story of display, which could also have involved body paint and clothing.

Only a handful of teeth remain of the inhabitants of Üçağizli from a presence lasting many millennia. One tooth is reportedly large, but overall they seem to represent *Homo sapiens*, suggesting that the occupation was indeed by modern humans. The food refuse left at the site shows that many large animals (such as wild goats and pigs, red, fallow, and roe deer, and wild cattle) and also small ones (hare, squirrel, partridge) were processed and consumed there. The diet was supplemented with shellfish, and at times even fish such as bream seem to have been eaten. Many of the stone tools must have been manufactured to kill or process game, and they include spear points and many narrow blades modified into knives, scrapers, and points. Tools and food-gathering behavior mostly show only gradual changes through the cave sequence, suggesting a presence over many millennia. There is one exception, about 41,000 years ago, however, when the inhabitants switched from using hard hammers (for example, cobbles) to soft hammers (probably of bone or antler) to make their tools, giving them greater control over fine working and shaping. Overall, there are signs that the occupations gradually became less episodic and longer in duration, and the food consumed more varied, perhaps indicating the growing abilities of the Üçağizli hunter-gatherers to adapt to their local environment.

As we saw, physical traces of the inhabitants of Üçağizli over a period of more than 10,000 years are few and far between, although they were almost certainly modern humans. To find further evidence of the people of this period, we need to travel south from the cave, first to Lebanon and then the Nile Valley. As mentioned, the Emiran and Ahmarian industries of Üçağizli resemble those found elsewhere in the Middle East, and one of the key sites, Ksar 'Akil, is in Lebanon, about 250 kilometers farther south. This rock shelter just outside Beirut contains nearly twenty meters of deposits, rich in fossils and artifacts. A Jesuit priest and archaeologist, Father J. Franklin Ewing, first excavated the site in the 1940s, and it has been reinvestigated intermittently since, when the political situation has allowed. The Emiran and Ahmarian levels date from around 42,000 to 35,000 years ago, and the animals hunted were comparable to those found at Üçağizli. There is another major similarity in the presence of many shell beads, but there is one significant extra at Ksar 'Akil: compared with the scattered teeth found at the Turkish site, Ewing discovered the partial skeleton of a child, who acquired the nickname "Egbert." Unfortunately, in the chaos that has intermittently descended on Lebanon, the original fossil vanished, hopefully only temporarily, but I studied a replica of the child's skull made by Ewing's team. It is undoubtedly a modern human child and surely shows us the species responsible for the shell beads in Turkey and Lebanon some 40,000 years ago.

There are further possible connections in the only North African fossil human of comparable age to Oase, Üçağizli, and Ksar 'Akil: the 40,000-year-old Nazlet Khater specimen from Egypt. This skeleton is of a young man who was deliberately buried in one of the oldest known mines, dug to extract chert rocks for toolmaking, close to Luxor on the Nile. His short and apparently well-muscled frame has many marks of wear and tear in one so young, leading to the suggestion that he might even have been an ancient slave forced to work in the mine; but if so, he seems to have been laid out in a decent burial. His skull is clearly that of a modern human, yet in the slightly retreating forehead, the shape of the face, and the wide ascending ramus of his jawbone, he resembles the Oase fossils. His teeth are not as large, and the poorly preserved inner surface of his jaw seems not to show an H-O foramen. But the tools made from the stones he must have quarried bear a general resemblance to those found in Lebanon, Turkey, and the Bohunician sites of central Europe. So perhaps this gives us a clue as to where the earliest modern people in Europe came from, and their route into Europe from the Middle East, during a brief warm phase about 43,000 years ago. Did people like those at Nazlet Khater and Oase, bearing proto-Bohunician artifacts, travel around the Turkish coastal plains (more extensive then because of lower sea levels) to the Black Sea, and then up the Danube corridor toward central Europe? If they did, it seems that their pioneering long trek might ultimately have ended in failure, and it was the succeeding Aurignacians who next took up the challenge of Europe and the Neanderthals.

Highly controversial evidence that those Aurignacians met (and maybe ate?) one of the last Neanderthals was published in 2009. The claim came from a detailed study of jaws and teeth found many years ago in Aurignacian levels at the cave of Les Rois in southwestern France. One child's jawbone is clearly modern, but the other shows possible Neanderthal features in its teeth, and growth lines on its teeth show a Neanderthal-like pattern of development (see chapter 3). Furthermore, while the more modern of the jawbones shows no signs of human modification, the Neanderthal-like one carries cut marks that suggest defleshing, and possible removal of the tongue. The authors were cautious about whether this represented cannibalism, and considered the alternatives that the cut marks were evidence of symbolic use of the child's remains as a trophy, or as postmortem treatment before burial. They concluded with three possible explanations for this unprecedented discovery: that the Neanderthal-like jawbone suggests symbolic use or consumption of a Neanderthal child by early Cro-Magnons; that Aurignacian tools were in fact produced by human groups bearing both modern and Neanderthal characteristics, that is, a mixed or hybrid population; or that all the remains from Les Rois represented modern humans, but that some displayed more primitive characteristics than normal in the Cro-Magnons.

Any one of those three explanations would be important for our understanding of events in Europe about 35,000 years ago, and the first two would be sensational evidence in support of one or the other of the two leading scenarios for the extinction of the Neanderthals that we discussed already: that they were replaced by the Cro-Magnons after a period of coexistence and possible interaction, which might have included direct competition between them; or that the Neanderthal and Cro-Magnon populations interbred and blended with each other during their possible coexistence. However, before getting carried away with these scenarios, we have to remember that the tantalizing second jawbone bearing cut marks is very incomplete, and the authors acknowledged that their identification of it as Neanderthal was tentative (other workers have identified the teeth in it as modern). New studies of the Les Rois fossils are planned, which could even include DNA investigations, and these should certainly help to clarify the picture. And because of the importance of Les Rois, new excavations are taking place there, and further human fossils have already been found. So hopefully this new evidence will help to solve the intriguing mystery of the children of Les Rois—their identity and their fates—in terms of events 35,000 years ago in Europe, and in terms of what they can tell us about human evolution in general. In that respect, it is also interesting to look at events at this time far across the huge landmass of Eurasia, in China.

Considering its vast size and its rich finds of *Homo erectus* fossils from sites such as Zhoukoudian ("Peking Man"), China has surprisingly little evidence of early modern humans. This may partly be a reflection of the relatively late arrival of *Homo sapiens* in China, but it's likely that the time scale for modern human evolution in China is, in fact, not much different from that of Europe, where by comparison there are dozens of Cro-Magnon skeletons and scores of rich sites representing the cultures of the Upper Paleolithic. To make matters worse, the richest collection of such material—from the Upper Cave at Zhoukoudian—was lost when it disappeared in 1941, following the Japanese occupation of Beijing.

However, the situation improved in 2003 with the discovery of a partial human skeleton in the nearby Tianyuan Cave, which was dated by radiocarbon to about 40,000 years. The cave contains remains of mammals such as deer and porcupine, and the animal bones in the level of the skeleton show signs of processing by humans, but no artifacts have been reported so far, and it's unclear how the well-preserved but incomplete skeleton got there: was it originally a burial, or did the individual concerned die in the cave? Thirty-four bones of the skeleton are represented, including the lower jaw, shoulder blades, various arm and leg bones, and parts of the hands and feet, and they are mostly medium-sized, making it difficult to judge the sex of the individual concerned, without a complete skull or the bones of the pelvis. However, he or she was at least middle-aged based on the high degree of tooth wear and the wear and tear on the bones preserved, including evidence of osteoarthritis.

Research by Hong Shang, Erik Trinkaus, and their colleagues shows that the Tianyuan individual had many typically modern features such as a well-developed chin and features of the shoulder blade, arm bones, and thighbone. But it also has relatively large front teeth, and one of the finger bones has an expanded tip, which is common in Neanderthals. However, the limbs are proportioned more like those of modern humans who derive from warm climates, rather than cold climates, and in this respect they resemble the early Cro-Magnons and not the Neanderthals. That fact leads to the likelihood that these early moderns, far from their warm African evolutionary homeland, were using cultural means to cope with the cold environments in which circumstances had landed them.

There are two well-known "rules" about the size and shape of organisms, related to their basic need to lose or conserve body heat, named after the biologists who first laid them out clearly. *Allen's rule* states that warm-blooded creatures in colder climates will generally have smaller limbs (or other appendages such as ears or tails) than their equivalents in warmer climates. The reasons for this may be both genetic (inherited) or acquired (for example, by the more restricted flow of growth nutrients to the extremities in cold conditions), but the result is a reduction of heat loss in cold conditions and an enhancement of it in hot conditions (think of the large ears of African elephants, for example). *Bergmann's rule* states that, in general, the body mass of organisms will increase with an increase in latitude or exposure to cold. This is related to the fact that a large animal will have a proportionately lower surface area for its weight than a small animal. Thus all other things being equal, a large animal will be better at retaining its body heat, while a smaller-bodied animal will be better at losing it. In humans, this may translate into overall build, with a more spherical shape (short and wide) performing better in cold conditions, while something more cylindrical (tall and narrow) would be favored in hot dry conditions, where heat loss is more important and increased surface area for sweating would be advantageous.

If we look at fossil humans, these rules generally work well. African *Homo erectus* and *heidelbergensis* bones suggest that the individuals concerned were lanky and long-limbed, while Chinese *erectus* and European *heidelbergensis* were relatively more compact and heavily built. When we get to Neanderthals, the pattern is even clearer, with skeletons from the last Ice Age in Europe being particularly stocky, with shorter extremities in the arms and legs. But some thirty years ago Erik Trinkaus noted that the Cro-Magnons who lived in much the same environments as Neanderthals were more like recent African populations in body shape than the Neanderthals. And the same thing now seems to apply to the earliest modern skeleton we have from the north of Ice Age China. Of course, there may be other factors at work in body size and shape, for example, the need for power, speed, or mobility, but it seems likely that human behavior, in the form of what is called *cultural buffering*, is also having an effect here.

Neanderthals are often considered to be a cold-adapted species, but in fact they ranged widely in time and space (for example, they are found alongside hippos and warm Mediterranean forests near Rome 120,000 years ago, and also with woolly mammoths in a bleak and seemingly treeless landscape in Norfolk about 60,000 years later). Archaeological data suggest that they actually shunned the coldest locations in Europe, especially those with the lowest winter temperatures and the highest wind chills, and this may be because, despite their physical adaptations, they lacked well-tailored clothing and well-insulated dwellings. These "luxuries" are actually essentials for survival in the cold, and we should remember that it's not just adult survival that is important here. Human babies are notoriously susceptible to low temperatures, because with their smaller body size they are less good at retaining heat (Bergmann's rule again), and their body's ability to stabilize temperature is still only developing. Apparently without sewing and weaving, the Neanderthals probably made fur wraps and ponchos, and fixed animal skins together with thongs or sinews, but we know from bone needles, impressions of weaving in clay, remains of huts and tents, sculptures, and the complex arrangements of beads and fastenings left from clothing in their burials that the

Cro-Magnons were able to provide much better insulation for their adults and children. Even the foot bones of the Cro-Magnons provide clues about another of their innovations: shoes.

Erik Trinkaus's anatomical study of the foot bones of modern people who regularly do or do not wear shoes showed that the difference is reflected in the robustness of their toe bones, since wearing shoes deflects some of the pressure otherwise applied to the middle of the foot during walking; hence some of the toe bones are less strongly built when shoes are worn. The distribution of accessories in Cro-Magnon graves suggests the original presence of clothing, and at Sungir in Russia, two adults and two children were buried with elaborate grave goods about 28,000 years ago. The burials were covered in hundreds of tiny mammoth ivory beads that must have been sewn on clothing which has since rotted away, and these are also abundant around the ankles and feet, suggesting the presence of decorated shoes or boots. Moreover, the foot bones of the otherwise very robust Sungir man show the telltale signs of gracility that indicate the wearing of shoes in people today. Such signs are not known in the toes of Neanderthals, nor in the much earlier modern humans who lived in Israel about 110,000 years ago. But they are there in the Tianyuan foot bones, which are over 10,000 years older than those from Sungir. Thus the earliest known modern skeleton from China seemingly had the benefit of shoes, which would have eased travel across difficult terrain and, if waterproof, could also have provided protection from the cold, wet, and snow. Although we don't know much about the way of life or culture of the isolated mystery *Homo sapiens* from Tianyuan Cave (but see chapter 3), the body of this man or woman tells us that increased protection from the Ice Age environment had arrived in northern China at least 40,000 years ago.

Let's now return to the ultimate ancestors of the early modern people we have been discussing from Europe and China, to fossils from Ethiopia. I explained in chapter 1 how the Omo Kibish fossils found by Richard Leakey's team in Ethiopia in 1967 were important to me in my formulation of a Recent African Origin model. Also, the dating of the fossils was not very secure at that time, and the specimens themselves were subject to interpretations very different from my own. In chapter 9 we will see how discoveries in the last decade have put these southern Ethiopian finds back in the spotlight, but now I want to focus on an arid region in the northern part of Ethiopia. Inland from the Horn of Africa, the triangular Afar depression is sinking as Africa splits apart along its great eastern rift, and the resultant basin has accumulated rich sediments through more than 5 million years of geological and early human history. The area has yielded finds of several australopithecine species including the 3-million-year-old remains of "Lucy," and younger fossils of *Homo erectus* and *heidelbergensis*. In 2003, the village of Herto became famous with the announcement of a rich site dating from about 160,000 years ago. The first clues came with the discovery of a hippo skull, evidence that this dry area once had fertile lakes and rivers, and, just as important, the skull showed butchery marks made by early humans. Systematic excavations by the paleoanthropologist Tim White and his colleagues uncovered other animal fossils, stone tools, and the remains of seven humans. Of these, one had a nearly complete adult skull, another was the braincase of a child of about six years at death, and a third represented part of another adult skull. All were very large in size. Bill Howells, mentioned in the first chapter, had spent half his life meticulously gathering data on modern human skulls to map the variation of our species, but the most complete Herto skull exceeded in major dimensions all 5,000 of those he had measured from around the world! Perhaps such size was necessary 160,000 years ago if hunting hippos was on the agenda. Even today these temperamental beasts are reputed to be the highest cause of human deaths among all of Africa's mammal fauna.

The brow ridge of the most complete skull is strong and projecting, above a broad and flat face, while the braincase is high, rounded, and modern-looking overall. The child's skull is too young to bear a brow ridge yet is equally modern-looking. But the rear portions of the two adult fossils are very strongly built, and reminiscent of the same regions in the Broken Hill skull from Zambia, which I classify in the species that was probably ancestral to *H. sapiens* in Africa: *H. heidelbergensis*. Two of the three Herto braincases also display evidence of human modification, including cut marks; could this be evidence of cannibalism? The scientists who described the material considered it was more likely that the skulls were kept as trophies or revered objects, because the child's skull was highly polished with extensive evidence of wear and scraping, as though it had been regularly handled, and thus it might have been linked with postmortem ritual behavior, perhaps even as a drinking cup—but this must remain speculation for now. It was hard to know how the Herto fossils should be classified. The authors of the *Nature* article of 2003 assigned them to a new form of *H. sapiens* called "*H. sapiens idaltu*" (*idaltu* means "elder" in the Afar language), because of their large size and robustness. I argued at the time—and still do—that the specimens are not so remarkable when compared with some other early moderns, such as those who lived in Australia near the end of the last Ice Age, so the distinct *idaltu* subspecies name is probably unnecessary. Crucially, however, the deposits containing the Herto fossils were sandwiched between volcanic layers dated by the argon-40/argon-39 method (see chapter 2) to 154,000 years old and 160,000 years old; therefore, along with the Omo Kibish 1 skeleton from southern Ethiopia, these may be the oldest definite traces of humans like us anywhere in the world.

In this chapter we have seen how recent discoveries opened up completely new windows on the evolution of *Homo sapiens*. They illuminated our African origins more than 150,000 years ago, they helped us map the dispersal of early moderns out of Africa and into Asia and Europe, and they told us much more about our cousins and possible competitors the Neanderthals. Fossils are vital to our story, of course, but so is the record of human behavior. However, it too requires careful extrapolation from the evidence left behind, as we shall see in the next two chapters.

5

Behaving in a Modern Way: Mind Reading and Symbols

I USED TO BELIEVE THAT HUMANS HALF A MILLION YEARS AGO were very different from us in their behavior. Even though they were making beautifully shaped handaxe tools, they were probably much closer to apes than they were to us, living a very basic life and getting what meat they could by scavenging from the kills of far better predators. But as I followed the progress of excavations at Boxgrove in Sussex in the 1990s, I started to realize that things were not so clear-cut, and I began to develop much greater respect for the abilities and achievements of those early Britons, living at the edge of the inhabited world 500,000 years ago. I well remember a conversation with Mark Roberts, the director of the Boxgrove excavations, where we speculated about how the Boxgrove people got their meat. They were using almond-shaped handaxes, skillfully made from flint, to skin, disarticulate, and butcher carcasses of horse, deer, and even formidable animals like rhinos. We know this because the animal bones, covered in impact and cut marks, are scattered across an ancient preserved landscape. So were the Boxgrove humans actively hunting game, even as large as rhinos, or were they scavenging already dead animals? Well, they were certainly spending a considerable amount of time on the butchery, seemingly getting at every bit of nutrition available, in a potentially dangerous open landscape. This indicates that they were organized enough to secure the carcasses from competing animals, such as lions, hyenas, and wolves, which we know were also at Boxgrove.

Wherever bones had both cut marks and signs of carnivore chewing, the cut marks were always made first—so these humans had primary access. And more direct evidence of hunting may come from an apparent spear-point hole in the shoulder blade of a horse. Although no spears are preserved in the conditions of the sediments at Boxgrove, wooden spears of yew and spruce have been discovered at Clacton in Essex and Schöningen in Germany, dating from 300,000 to 400,000 years ago. The Clacton "spear" is only a broken tip, but the German spears are some two meters long and beautifully made, and their use in hunting seems established by the fact that they were found among twenty or so horse skeletons. Archaeologists are still debating whether such spears were for throwing or thrusting but, in either case, these ancient people were clearly capable of tackling large and dangerous wild animals for their next meal.

How might relatively puny humans, albeit ones who were almost certainly stronger, fitter, and more muscular than the average person today, have coped with such dangerous predators and prey? Rocks, sharp stones, and wooden spears must have been part of the answer for a creature that was not equipped with speed, great strength, or sharp teeth or claws, but cooperation and cunning were probably even more important. Mark Roberts told me of a conversation he had had with an expert on wild rhinos in Africa, when he asked how, armed only with wooden spears, a man could kill a rhino. Well, the expert said, he would never be so foolish as to attempt such a thing, but when Mark pressed him further he said if he *really* had to do it, he and some friends would wait to find a solitary rhino asleep in the shade of a tree. Then, spears at the ready, they would creep up on the rhino, stab it quickly in its exposed belly as it slept, and hurriedly climb the tree. They would then hope the rhino bled to death; otherwise they could be stuck up the tree for a long time!

But at least *four* rhinos were butchered at Boxgrove, over an unknown period of time, suggesting this was not an exceptional event. Instead of a one-off foolhardy or lucky enterprise, it looks far more likely that this was part of the normal repertoire of *Homo heidelbergensis*. No doubt the ability to outthink and outwit the opposition, to "predict" its likely behavior, and the behavior of your fellow hunters, would have been crucial. This mind-reading ability, first developed in our primate ancestors, is now a significant part of the characteristics that have made humans, and particularly modern humans, so special. For some experts, this led to a heightened ability to control thoughts, emotions, and actions; to plan far into the future; and to evolve self-consciousness. Through growing social complexity we also developed greater powers of imitation, social learning, imagination and creativity, cooperation and altruism, enhanced memory, and complex language.

So far in this consideration of modern human origins I have focused largely on the physical evidence of what it is to be a modern human—for example, features in the skull, jaws, and body that survive in fossils—and what they tell us about how we may have evolved. But of course so much of what we think of as human lies in our behavior, many aspects of which are accentuated versions of what we can find in our closest living relatives, the great apes—things like tool manufacture and use, a long period of infant dependency, and social complexity. Other aspects seem to be quite unique to us among the primates—things like composite tools, art and symbolism, elaborate rituals and religious beliefs, and complex language. The gap between us and the great apes may seem more like a vast chasm, but we are the only surviving representative of what were extensive evolutionary experiments in becoming human, and so many of the features we think of as unique to us were shared, to a greater or lesser extent, with now-extinct species like *Homo erectus* and the Neanderthals.

There are certainly hints in our biology of odd quirks that, if we understood them better, could give us clues to how humans came to be so different, or at least so much more complex socially, than our primate relatives. For example, in most of the primates—and probably our ancient African ancestors—the outer covering of the eyeball, the sclera, is dark brown. This means that the pupil and iris in the center of the eye, which move to focus the gaze, are difficult to differentiate from the surrounding tissue, especially where they are dark. But humans have an enlarged, unpigmented, and therefore white sclera, which means we can detect where other people are looking; equally, they can detect where we are looking. This must have evolved as part of the development of our social signaling, enabling us to "mind-read" each other. (This idea even has a name: the Cooperative Eye Hypothesis!) Similarly, many domestic dogs have an accentuated white sclera compared with their wild wolf ancestors, which perhaps evolved to augment the close social relationship between dogs and humans.

Another remarkable feature in humans is the large size of the penis, of which much was made when Desmond Morris's book *The Naked Ape* was published in 1967. In fact, the human penis, when erect, is no longer than that of chimpanzees and bonobos, although all of these are about double the length of the penis in much larger orang and gorilla males. But the human penis is considerably thicker than any of the others and has a much more bulbous end. Explaining how and why these differences evolved has led to much speculation, ranging from the enhancement of pleasure to displacing the sperm of competing males, to providing a very obvious sexual display as a signal to either females or other males. The other obvious external part of the male reproductive organs—the testicles, containing the sperm-bearing testes—is less distinct in humans, intermediate in size between that of chimps (very large) and orangs (small) and gorillas (tiny). It is believed that this is related to both frequency of mating (high in chimps, low in gorillas) and competition between males for impregnation of fertile females (again high in chimps, low in gorillas). Humans thus fall between the extremes, suggesting that we mate (or, more appropriately, our ancestors mated) fairly often, but with only moderate levels of promiscuity compared with chimps.

Darwin had to make extensive use of analogies with other animals, because the fossil and archaeological evidence that he would have valued so much took many more years to blossom. However, accepting our close kinship with the great apes, he recognized similarities between their behavior and intelligence and ours. In 1871 he wrote:

As man possesses the same senses with the lower animals, his fundamental intuitions must be the same . . . But man, perhaps, has somewhat fewer instincts than those possessed by the animals which come next to him in the series. The orang in the Eastern islands, and the chimpanzee in Africa, build platforms on which they sleep; and, as both species follow the same habit, it might be argued that this was due to instinct, but we cannot feel sure that it is not the result of both animals having similar wants and possessing similar powers of reasoning. These apes, as we may assume, avoid the many poisonous fruits of the tropics, and . . . we cannot feel sure that the apes do not learn from their own experience or from that of their parents what fruits to select.

Darwin has been criticized for his excessive anthropomorphism in the recognition of "human" behavior in other animals, and given his lack of reliable data on great ape behavior—much of it based on captive animals or the tales of explorers—it is not surprising that he got things wrong at times. But overall he was cautious in his extrapolations. We now know far more about our close evolutionary relationship to our primate kin, and we should not be surprised to find both shared behaviors and shared brain pathways behind them. Thus monkeys and apes can recognize the different elements and expressions that make up faces from simple drawings rather than accurate pictures. The neuroscientist Vilayanur Ramachandran stressed the potential importance of *mirror neurons* in their brains and ours, nerve cells that are triggered both when an animal performs an action and when an animal observes another animal performing that same action. Such acting out of deeds in the brain is thought to be important in human learning, social interaction, and empathy, giving primates the basic elements of "mind reading," which, as we will see, is so important in complex societies like ours.

But we do have a major problem when we turn to reconstructing the complexities of past human behavior, since what is left behind as physical evidence in the form of stone tools and butchered bones only represents the end products of chains of thoughts and deeds that are lost to us now, and which we attempt to reconstruct at our peril. Certainly we can turn to living apes to provide models for early human behavior for such activities as simple toolmaking and primitive hunting, but how much like a chimpanzee were, say, the *Homo heidelbergensis* people of Boxgrove in England 500,000 years ago, who were already living far from their tropical African homeland, making complex tools like handaxes, and acquiring not just small mammals but potentially dangerous big game such as horse, deer, and rhino? Just as significant, *H. heidelbergensis* already had a big brain, one nearly as large as ours today. To understand the evolution of those large human brains, we need to look at what they might have been used for.

There is now evidence that chimps in the wild do have "cultures," shared traditions of how to behave—for example, in gathering or processing food with tools—which differ from one group or regional population to another. These cultural norms are learned as the chimp grows up in its group, and female chimps seem to be prominent, both in passing on traditions to new generations and in developing new ones. Yet these cultures are still rudimentary, and chimps are seemingly a long way from the cultural repertoire of even the earliest humans in Africa 2 million years ago. We remain unique in the extent to which we modify the world we live in through the things we create. Beyond that, we create imaginary worlds that are entirely virtual, made up of thoughts and ideas—worlds that live in our minds, from stories and spiritual domains through to theories and mathematical concepts. Chimps possess basic concepts of cause and effect; for example, if they strip a grass stem and lick it, it will then be thin enough and sticky enough to be used as a probe to catch termites. But humans have the ability to imagine a much longer chain of cause and effect, to consider several different outcomes that could result from an action or an alternative action. Through the medium of language, we can communicate these complex concepts to each other, both those relating to the material world, such as how to make a fire, and those relating to imagined worlds, such as what may happen to us after we die.

Instead, should we perhaps turn to modern hunter-gatherers in places like Brazil, Australia, and Namibia to help us reconstruct how the Boxgrove people, or the Neanderthals, or our African ancestors lived? We have to use such data cautiously and always be aware of the assumptions and extrapolations we make, since much has evolved and changed in the intervening millennia. So how could such complexity of behavior, including the ability to create virtual worlds, have evolved? One possibility is that an increase in meat eating in our ancestors not only gave access to more concentrated foods, removing previous constraints on large, energetically demanding brains, but also set in motion far-reaching changes in behavior, enhancing the power of reading the minds not only of our prey but of members of our own social group.

Daily life in primate troops in the wild has been compared with the worst aspects of television soap operas or reality TV shows like *Big Brother*: bullying and domination by the strongest, fear and abuse for the weakest. Yet primate groups also demonstrate tenderness and affection, strong alliances for the greater good, and social bonds that can last throughout life. This brings us to what is called the Social Brain Hypothesis (SBH), advanced by psychologists and anthropologists such as Nicholas

Humphrey, Robin Dunbar, Richard Byrne, and Andrew Whiten. By this hypothesis our large brains have evolved not just in response to human needs for things like foraging and hunting skills, toolmaking, and invention, but also because of the complex societies in which we live. All primates have large brains for their body sizes relative to the average in mammals, particularly the so-called higher primates, the monkeys and apes. Brains are very demanding of energy; in fact, in humans, the demands of the brain are second only to those of the heart. So why would, say, a lemur or bush baby need a relatively larger brain than a hedgehog or a squirrel? One argument has been that the forested environments in which primates generally live require a keener intelligence to cope with problems, while another perspective has to do with the longer growth and development found in primates both before and after birth. Yet these explanations on their own do not seem sufficient, which is why SBH has gathered an increasing number of influential supporters.

Various comparative studies have shown that the relative size of the neocortex of the primate brain is much larger than normal in mammals (and in humans it constitutes a whopping 85 percent of total brain weight). The neocortex is part of the cerebral cortex and is known from brain mapping to be responsible for higher-level cognitive functions, such as learning, memory, and complex thought. That might seem to support the idea that it is large because the environment in which primates have evolved demands a keen intelligence in finding food and escaping predators. Yet plots of neocortex size against environmental complexity seem to explain less of the primate pattern than do plots against variables that reflect social complexity, such as group size, numbers of females in a group, frequency of social alliances, and amounts of social play, manipulation, and learning. Thus while environmental hypotheses tend to assume that animals solve problems individually by trial-and-error learning, without relying on the social groups in which they live, SBH proposes that such problems are solved socially, with the need for a larger neocortex to enhance social comprehension and cohesion. No doubt, in reality, both the social and general environment play a part in generating evolutionary demands, but in the case of humans it is very difficult to argue that general environmental demands have been anywhere near as important as social ones in the development of our extraordinarily large brains.

There is evidence that pair-bonded birds and mammals have relatively larger neocortices, and one possibility is that in higher primates, and particularly in humans, the social and thinking skills used in pair-bonded relationships have been extended many times over for creating and maintaining relationships between individuals who are not partners in reproductive terms. Thus individuals of the same or opposite sex may form bonds as intense and long-lasting as those normally found between pair-bonded mates in other species; in other words, humans have mates beyond just sexual mates. For this to work successfully and in the long term, it requires the extension of high-level social skills of trust, empathy, and synchronization of actions beyond the immediate "family group" and into the larger social community. These links would be valuable in terms of reliable support in times of difficulty, and through an extended web of such relationships crosscutting through the whole group, all could benefit from coordinated action, food sharing, protection from predators, et cetera.

Thus many scientists believe that our substantial brains evolved via selection for life in large groups, and this led to the development of deep social minds in primates, with the ability to "mind-read" (observe and interpret the actions of) others in the group, to learn and pass on "cultural" behavior within the group, and to cooperate not only for mutual benefit but for the benefit of others in the group. Mind reading, or possessing a "theory of mind" about oneself and others, can occur at several levels and for many different social purposes, for example, interpreting what individual A thinks about individual B and then behaving so as to manipulate the behavior of A toward B. (This social "skill" is sometimes known as *Machiavellian intelligence*, a term introduced by Byrne and Whiten, after the Florentine political philosopher Niccolò Machiavelli.)

Mammals and birds seem to have a first order of intentionality, that is, they are aware of their own behavior and its possible impact on others; as mentioned, some of this may be related to the demands of strong pair-bonding or living in herds or flocks. But by the age of four, human children can operate at two levels of intentionality in their social perceptions, that is, perceiving and interpreting not only their own behaviors but also the behaviors of those immediately around them. Thus children have the ability to recognize that others may have the same, or different, perceptions of the world compared with their own. At this stage such recognition means that they can begin to manipulate, or try to manipulate, those around them, whether these are parents, siblings, their peers, or their nursery school teachers. There is evidence that chimpanzees approach the same level as four-year-old children in their theory of mind, but they never move beyond this, whereas most humans develop further to cope with several higher levels of intentionality. Robin Dunbar illustrated this point with reference to Shakespeare's play *Othello,* where the playwright had to simultaneously handle four mind-states on the stage: Iago intends that Othello should believe that Desdemona loves Cassio and Cassio loves her. But Shakespeare moved beyond that because, to be successful, he also had to be able to visualize the audience's reaction to what he was writing— and so he was working to at least a fifth-order intentionality, right at the limits of human mind-reading abilities. These highest levels, supporters of the SBH argue, are unique to modern humans, and they evolved through the need for our ancestors to map the growing complexities of their social relationships. This in turn raises the question of why such complexities had developed.

SBH perhaps helps to explain something that does differentiate most human hunter-gatherer groups both from our primate relatives and from modern industrialized societies: egalitarianism. Hunter-gatherers usually have little in the way of owned material possessions, since they are difficult to maintain and transport with a nomadic lifestyle, and this social equality is reflected in things like food sharing, lack of formal leadership, and the prevalence of monogamous relationships. This last contrasts with the polygamy that characterizes primates such as baboons and gorillas, as well as many agricultural and pastoralist societies, where a few men may accrue disproportionate wealth, status—and wives. Maintaining social equality often requires positive coordinated efforts by the group to resist those individuals who try to assert excessive dominance. Coordination of activities extends to bands of women who plan ahead to go foraging for plant foods, insects, and small game, and to hunting bands, who must also plan ahead, communicate about tracks and signs, and adopt specific roles in catching and processing prey. In terms of the vital activity of food acquisition, the degree of coordination that a sophisticated social brain can help to deliver means that the group acts more like a food-gathering machine than the host of individual and "selfish" foragers typical of a monkey or ape troop.

But there are real practical limits on the size of a group that can successfully interact and function at a personal level, and this has been enshrined as *Dunbar's number*, following Robin Dunbar's research. In primates this may be the subgroup in a troop who regularly interact by means such as mutual grooming of fur, and can range up to about sixty. For Dunbar, what ultimately limits the size of such groups in different primates is the relative size of the neocortex, which governs how many friendly relationships or meaningful acquaintances can be successfully managed at any one time (although recent research suggests that the small regions known as the amygdala, located near the base of the brain, also play an important role in humans). Dunbar's number in modern humans seems to fall between about 100 and 220 (average 148), and the figure matches quite well with the optimum size of large hunter-gatherer aggregations, tribal villages, Hutterite settlements, small military units, and even the average number of people in effective social networks on the Web. As we shall see later in this chapter and the next, the relatively large size of human aggregations has had consequences in terms of the need to develop new means of communication (symbolism and language) and the evolution of more complex social structures and fully human culture.

With our large brains, evolved for and geared to interact flexibly with a network of people in our social group, we can exchange information of mutual benefit. But how free are we really in these interactions, and how important is our genetic inheritance in determining what we can and cannot do? In chapter 7 I will discuss our DNA and genes, and their importance in reconstructing the processes of human evolution, but there is no doubt that our DNA does provide a basic template for our behavior. It's a template that provides both limiting and varying factors in what we can and cannot do (for example, in determining the basic size and shape of our brain, the extent of human dexterity, running speed, acuity of vision and hearing). At the same time, it is obvious that humans can improve elements of performance through learning and practice, and many of these are influenced by differences in both the physical and social environments (for example, diet, health, upbringing, and social norms). So our DNA is more like a flexible container than a mold in the way it determines and sets limits on how we behave. Nevertheless, as we shall see, some scientists believe not only that the structure of the modern human brain is quantitatively different from that of earlier humans in its size and the extent of its gray matter, but that genetic changes unique to modern humans also qualitatively rewired our brains about 50,000 years ago, making us behaviorally modern at a stroke. If that was so, then despite their large brains, Neanderthals were fundamentally unlike us in their lack of human behavior, because they followed a separate evolutionary trajectory. That same lack would have applied to the modern humans who occupied Africa before 50,000 years ago, because they lived before the mutations occurred that made us fully modern.

How much of our modern behavior was shared by earlier human species? That is a very difficult question to address, let alone answer, from the data we have. Some researchers produced a sort of checklist of modern behavior, which, it is argued, characterizes humans today and which can then be used to examine the archaeological record for when and where those traits first appeared. Other workers would dispute their usefulness, their universality, and how accurately they can be inferred from the imperfect materials that survive in the ground from ancient times. But the list of behaviors often includes: complex tools, the styles of which may change rapidly through time and space; formal artifacts shaped from bone, ivory, antler, shell, and similar materials; art, including abstract and figurative symbols; structures such as tents or huts for living or working that are organized for different activities (such as toolmaking, food preparation, sleeping, and for hearths); long-distance transport of valued materials such as stone, shells, beads, amber; ceremonies or rituals, which may include art, structures, or complex treatment of the dead; increased cultural "buffering" to adapt to more extreme environments such as deserts or cold steppes; greater complexity of food-gathering and food-processing procedures, such as the use of nets, traps, fishing gear, and complex cooking; and higher population densities approaching those of modern hunter-gatherers.

In the last fifteen years there have been some remarkable discoveries that emphasize the complexity of early modern human behavior in Europe and elsewhere. Can any of these provide clues as to what triggered the changes that led to our modern minds and patterns of behavior? To understand this, we need to address the important question of symbolism, which many people think is a key to understanding what made us different from any creatures that came before us. We use symbols in so many ways today that they are a part of our lives, taken for granted, but without which we could hardly function. They may be pictorial, such as a drawing which, though small, black, and two-dimensional, resembles the larger solid object it represents—say, a picture of a stick-man or stick-woman, or a plane. Or it could be a written word, which does not resemble at all the object or action that it stands for—like the word *computer*. Or perhaps it could be a series of musical notes, which can only transform into sounds when interpreted through the eyes, brain, and actions of a musician. Symbols may equally represent social signals about group identity, wealth, or status; here I don't just mean bundles of banknotes, which are, of course, merely pieces of paper with some printing on them. Equally symbolic are fashions in clothing, hairstyles, cosmetics, tattoos, and necklaces, which are used to send signals about the wearer to other individuals in the group.

Symbols are meant to transmit messages and usually require the recipient to be able to interpret them in the same terms as they were originally intended, although they can, of course, equally be used to exclude those who cannot receive and interpret them—a secret code or a ritual handshake, for example, the meaning of which is known only to members of a privileged group. The archaeologist Clive Gamble, following the primatologist Lars Rodseth, emphasized that one of the most distinct things about modern humans compared with our primate relatives is our "release from proximity." Probably all humans before us, like the apes they evolved from, could only work through face-to-face encounters, but with the rise of symbolism (and the associated development of language), people were liberated from proximity and could communicate through time and space. Thus a message with social meaning could be transmitted remotely on behalf of a person or group, provided the recipients could decode it in the same terms.

About twenty-five years ago, the concept of a 35,000-year-old Human Revolution emerged. This revolution considerably preceded, but matched in significance, other transformations such as the domestication of plants and animals, the discovery of metalworking, and the development of industrialization. This revolution seemed to be associated with the arrival of the modern-looking Cro-Magnons in Europe; they had imported or rapidly developed a package of modern human behaviors, such as complex language, art, and specialized technologies, enabling them to replace the behaviorally inferior Neanderthals. In 1987 I was fortunate to co-organize an influential conference in Cambridge that became known as the "Human Revolution" meeting, where this view really took off. Many have since criticized the Eurocentric focus of those debates about the origins of symbolism and modernity. Certainly the European evidence for early symbolism still dominates the Paleolithic record, not only because of the large number of accessible sites that preserve the finds, but also because of the large number of trained archaeologists who are working on those sites. Thus Europe continues to surprise us in the richness of its record: so let's now look at some old and new finds that illustrate this.

Shortly before Christmas in 1994, three cavers decided to explore a small opening in the limestone cliffs in the Ardèche region of southern France. Clearing rock falls, they reached a shaft and, descending with a caving ladder, entered a huge chamber, richly adorned with stalactites. Moving on to another large chamber, they started to see fossil animal bones on the floor and, after some time, decided to head back. Quite by chance, the beam of one of their lamps fell on a small red ocher drawing of a mammoth; at that moment they knew they were on the brink of a great discovery. They were astounded at the galleries of hundreds of paintings and engravings in charcoal and red ocher that they and the prehistorians who joined them over the next few weeks were to discover in the cave now named after one of them: Chauvet. About 36,000 years ago, several artists must have walked deep into the cave with flaming torches and squatted down to draw two rhinos locked in combat in charcoal. Just to the left they then outlined the heads of three aurochs (wild cattle), and, in the center, one of them picked out outlines in the rock and drew four beautiful horses' heads. Each of the horse's heads seems to show a different mood or character, as though depicting the passage of time or some inner narrative.

In the year 2000, six years after the discoveries at Chauvet, an amateur caver found another long series of galleries not far from the famous painted caves of Lascaux, and this time the site was dominated by engravings of bison, horse, rhino, mammoth, and deer, with some human figures. This site at Cussac may be 10,000 years younger than Chauvet (dating work is ongoing), but its importance is increased by the discovery, below some of the decorated walls, of seven human burials, tucked into "sleeping nests" made in the floor of the cave by bears.

The Chauvet and Cussac art mirrors the much earlier find, which, in the face of tremendous opposition from the archaeological establishment of the time, finally proved that cave paintings really were the work of "Stone Age savages," as they were then termed. In 1879 the Spanish nobleman and amateur prehistorian Marcelino Sanz de Sautuola and his nine-year-old daughter Maria climbed up to investigate the small opening of a cave in the Altamira hills, near Santander in northern Spain. After entering, de Sautuola decided to carry out a small excavation; Maria, bored, wandered off with a lamp to explore. Moments later she cried out to him, "Papa, look—oxen!" She was probably the first human to have set eyes on the beautiful painted galleries of Altamira Cave for 13,000 years, but for de Sautuola this encounter was to lead to a nightmare of scorn and rejection, as he tried to get the results of his research in the cave accepted as evidence that the art was indeed Paleolithic, and not recently forged. Sadly, he died in 1888, fourteen years before his arguments were finally vindicated.

Once cave art like that from Altamira was accepted as the work of the ancient Cro-Magnons, serious debate soon followed about its meaning. The art mainly consisted of representations of Ice Age animals, dominated by bison, horse, auroch (as seen by the young Maria), and deer, but there were also rarer, less elaborate, depictions of people, as well as patterned or abstract designs. Early explanations for the art ranged from "art for art's sake," to animal worship, to hunting magic. Later, more elaborate theories suggested that the components of the art were patterned like a code and were symbolic of such things as maleness, femaleness, conflict, and death.

Going back to the time of decorated caves like Chauvet and Cussac, but farther north in what is now Germany, somebody carved an extraordinary object from one of the most recalcitrant of materials: a mammoth tusk. The object, about six centimeters long and three centimeters deep and wide, is a statuette of a large-shouldered and generously proportioned woman—but who completely lacks her head. The head is not missing because of damage; it was never there, and just off-center from where the neck would have been, the carver instead made an ivory ring from which it must have been suspended as an amulet. Because of its large breasts and clearly marked vulva, the statuette has even been called pornographic, but it seems to my eyes much more about fertility, with its prominent belly and breasts perhaps swollen with milk, than pure sex. The cave of Hohle Fels is one of four in a region near the Danube that has now produced over twenty ivory figurines of horse, mammoth, lion, bison, and bird, as well as two statuettes, one tiny and one about thirty centimeters long, representing male human bodies with a lion's head. There are also many hundreds of ivory beads, which must have been strung as pendants, necklaces, or bracelets. All required skillful manufacture and probably date from 35,000 to 40,000 years old.

Three of the German sites have also produced the oldest-known musical instruments: four flutes made from the perforated wing bones of swans and vultures, and four manufactured from carefully refitted segments of mammoth ivory. The most complete would have originally been about thirty-four centimeters long and was made from the radius of a massive griffon vulture, in which five finger holes had been carved with stone tools. At the playing end, the maker also carved two deep notches, into which the musician would have blown. This flute was found only seventy centimeters from the astonishing ivory figurine described earlier, although it is not possible from the cave sediments to know how closely they were associated in the lives of the people who made and used them; they could have been exactly contemporary or centuries apart. Even more impressive in terms of manufacturing skills are the smaller surviving fragments of flutes made from ivory, where the curved form of the mammoth tusks had to be worked into long straight segments, which were drilled with playing holes and then fitted back together precisely with airtight seals. A modern replica of one of the bird-bone flutes with only three holes in it shows that it could produce four notes and three more overtones, so it seems that all of these instruments would have had the range of modern equivalents, though between them they would have differed in pitch.

Not only are these skillfully made flutes the world's oldest known

Objects from Hohle Fels Cave, Germany: (*top left*) "Lion Man"; (*top right*) flute; (*bottom left*) the astonishing "headless Venus"; (*bottom right*) waterbird.

musical instruments, but they and the female figurine also date from the very beginning of the Aurignacian occupation in these German caves, implying that such traditions must go back even farther in time, in Europe or in an even earlier homeland. There seems little doubt that they and the paintings in sites like Chauvet are the handiwork of modern humans, although Neanderthals must still have survived in some parts of Europe at this time. There has been much debate about how modern the Aurignacians actually were in their behavior, with some experts doubting the great ages assigned to the Chauvet art and the ivory statuettes from Germany. Others have even speculated that it could be the handiwork of Neanderthals or of a hybrid population, citing claims for an even older flute in a Neanderthal cave called Divje Babe in Slovenia. There, dating from perhaps 50,000 years ago, a cave bear femur was discovered with two clear holes in it, and possible signs of two further holes, where it was broken at each end. However, three separate studies showed that the bone was gnawed by carnivores at each end, and it was argued that the holes did not demonstrate human workmanship and were more likely to have been punctures from the canine tooth of a large mammal, such as a cave bear or wolf. Arguments continue about this object, but until further confirmatory finds are made in Neanderthal occupation sites, I do not think we yet have evidence that Neanderthals made music—though some reconstructions of their vocal tract suggest they might have been good, if rather high-pitched, singers!

Music in the form of singing and clapping seems to be universal among humans today, even if accompanying instruments may be as simple as tree-trunk drums, rattles, or sticks that are banged together. In *The Descent of Man*, Darwin was puzzled by this universality, saying, "As neither the enjoyment nor the capacity of producing musical notes are faculties of the least use to man in reference to his daily habits of life, they must be ranked amongst the most mysterious with which he is endowed." For some scientists, music is just a by-product of our language capacity and our ability to recognize patterns even in sounds such as a howling wind, running water, or human chanting. For others, despite Darwin's negative views about its usefulness, music is closely linked to the evolution of language and of complex modern human societies, where it would have played a critical role in cementing social relationships and in group rituals and ceremonies. In conveying meaning, music as a form of communication would then have formed an important part of the symbolic revolution. Its importance to humans seems to be confirmed by neuro-imaging of the human brain, where areas of importance in lan-

Excavations in deposits outside Vogelherd Cave, Germany,
which have produced several Aurignacian figurines.

guage, memory, and emotion are activated, and endorphins—feel-good
hormones—are released.

To many researchers in the last century, the astonishing evidence of
the complexity of Cro-Magnon behavior in Europe seemed to materialize
out of nowhere—there were no antecedents anywhere else. Even in the
late 1980s and 1990s many people, including me, seriously considered the
idea that there could have been a sudden origin of the suite of modern
features, both physical and behavioral, but it was unclear whether these
coincided or were separated in time by as much as 100,000 years. As data
continued to emerge from Africa that modern humans had indeed originated there, for many archaeologists the concept of a Human Revolution
moved there too, and the timing of the revolution was moved back with
new dating to the beginning of the African Later Stone Age, about 45,000
years ago. That is somewhat of a minority view now, although espoused
by the archaeologist Richard Klein. In his opinion, about 50,000 years
ago there were mutations in African early modern humans that enhanced

brain functions, producing changes in cognition or language. In turn, these changes would have generated new opportunities for further behavioral changes or innovations, which would have catalyzed the emergence of the fully modern pattern through feedback effects, eventually settling into the pattern we recognize today as behavioral modernity. This reshaping led to the successful expansion of modern humans and now-modern behavior beyond Africa, and the replacement of remaining archaic populations such as the Neanderthals. Thus morphological evolution and behavioral evolution were decoupled, since morphological modernity evolved before behavioral modernity.

This pattern is counterintuitive for people who favor the idea that behavioral changes lay behind the transformation of the archaic skeletal pattern into that of modern humans, where the use of increasingly sophisticated tools removed the necessity to maintain the robust bodies of our ancestors. If that were so, the behavioral changes should have preceded the physical ones, not the other way around, since they were driving the process of modern human evolution. However, Klein's view is based on the fact that, despite their morphological "modernity," 100,000-year-old fossil samples from sites such as Klasies River Mouth in Africa and Skhul or Qafzeh in Israel are associated with Middle Paleolithic artifacts, in many ways comparable with those made by Neanderthals. And, according to Klein, they apparently lacked many other aspects of "modern" behavior as well, though, as we shall see, that viewpoint is increasingly under attack. Klein argues that after about 50,000 years ago modern human morphology essentially stopped evolving, while cultural changes accelerated rapidly, even exponentially, from that point on.

Brain size had achieved essentially modern levels at least 200,000 years ago and was actually larger in the earliest modern humans and the late Neanderthals than the average today (though we need to remember that their bodies were also somewhat heavier and more muscular). So, Klein argues, the brain changes that occurred around 50,000 years ago must have been in organization, not size—something which we are very unlikely to pick up from the fossil evidence. However, ongoing work on modern human and Neanderthal DNA might eventually be able to compare elements of brain function in these two species, although we are unlikely ever to have well-enough-preserved DNA from our ancient African ancestors for such studies in their case.

A more gradual "revolution" position is now held by one of the people who was in the vanguard of the debates we had in Cambridge in 1987: my conference co-organizer Paul Mellars. He argues for a period of accelerated change in Africa between about 60,000 and 80,000 years ago, as shown by the following developments recorded in South African cave sites: new and better-controlled techniques for producing long thin flakes of stone blades; specialized tools called end scrapers and burins, which were probably used for working skins and bones; the production of tiny stone segments that must have been mounted on handles of wood or bone to make composite tools; complexly shaped stone tools such as "leaf points"; relatively complex bone tools; marine shells perforated to make necklaces or bracelets; red ocher (natural iron oxide) engraved with geometric designs suggesting early artwork; greater permanence and differentiated occupation areas in caves; new subsistence practices such as the exploitation of marine fish as well as shellfish; and perhaps intentional burning of undergrowth to encourage the growth of underground plant resources such as tubers. Mellars suggests that a neurological switch to modernity in the brain, alongside rapid climatic fluctuations, could have been the driving forces behind this period of heightened cultural innovations, of which the Toba eruption of about 73,000 years ago might have been a part.

The concept of a Human Revolution in the arrival of behavioral modernity during the last 50,000 years, as argued by Klein, or between 60,000 and 80,000 years, as argued by Mellars, has been strongly challenged by archaeologists like Sally McBrearty and Alison Brooks, who consider that such views display a Eurocentric bias, even if the original Human Revolution model has now been transferred to Africa. This is because by focusing on changes that occurred at the Middle Paleolithic/Upper Paleolithic or Middle Stone Age/Later Stone Age transitions (in Europe and Africa, respectively), there is a failure to appreciate the depth and breadth of the African Middle Stone Age record that preceded the time of the supposed revolution by at least 100,000 years. In their view, "modern" features such as advanced technologies, increased geographic range, specialized hunting, fishing, and shell-fishing, long-distance trade, and the symbolic use of pigments had already developed in a broad range of Middle Stone Age industries right across Africa between 100,000 and 250,000 years ago. This suggested to them that an early assembly of the package of modern human behaviors occurred in Africa, followed by much later export to the rest of the world.

Thus the origin of our species, both behaviorally and morphologically, was linked to early developments in Middle Stone Age technology, and not to changes that occurred much later, toward the end of the Middle Stone Age. McBrearty and Brooks also pointed out that by placing the most important changes close to the time of the exodus of modern humans from Africa, there might be an inference that those changes were the necessary ones to enable humans to exit from Africa and thrive beyond there, further implying that those who were left behind in Africa were in something of a cultural backwater. And as for the whole "revolution" concept, McBrearty said that "this quest for this 'eureka moment' reveals a great deal about the needs, desires, and aspirations of archaeologists, but obscures rather than illuminates events in the past. It continues to put Europe on centre stage, casting it either as the arena where the actual events of human origins were enacted, or as the yardstick by which human accomplishments elsewhere must be measured."

We do need to find the earliest evidence for symbolic behavior in the archaeological record—a key factor in resolving this debate—and whether it ever extended beyond our species. This brings us to the critical question of how to recognize symbolism, when we cannot mind-read the intentionality of people in the distant past. I regularly receive correspondence, pictures, and e-mails from people who are convinced that stones they discovered had been shaped by ancient humans into depictions of animals or human faces, and yet these are almost certainly natural objects, shaped by geology rather than a premodern human species. Our brains and eyes have evolved to recognize patterns, so a pebble with two round grooves and one straight one may look like a face to us, even if the holes are entirely natural and the stone came from deposits that are millions of years old. Some people consider that handaxes a million years old are symbolic objects, since they are generally shaped to look symmetrical and seem overdesigned for their function, if this was primarily to butcher ani-

mal carcasses; thus they may have had a social as well as a functional purpose. A famous paleoanthropologist reportedly said to the archaeologist
Desmond Clark that handaxes were so sophisticated they indicated that
Homo erectus must have had language. But Clark, noting that handaxes
had hardly changed shape through a million years of the Lower Paleolithic
(Lower Old Stone Age) and across three continents, answered that if that
were so, these ancient people were saying the same thing to each other,
over and over and over again!

As we move on in time to about 300,000 years ago, the more complex
technologies that are associated with the Middle Paleolithic, made both
by Neanderthals and by the lineage of *Homo sapiens* in Africa, start to
appear. Techniques that required more distinct steps in the manufacture
of tools became widespread across Africa and western Eurasia, and the
first truly composite tools appeared. Even before this, the Schöningen
spears showed ancient human preplanning that must have happened in
several stages over more than one day, and by 260,000 years ago, early
humans at Twin Rivers (Zambia) were apparently manufacturing stone
segments and points designed to be mounted on wooden handles. They
also left behind lumps of natural pigments in many colors, some locally
derived and others collected from a distance. Hematite (red iron oxide)
could have had functional as well as symbolic uses, such as treating animal
skins, as part of adhesives for composite tools and even as an insect repellent. But it is possible, although unsupported by other evidence, that the
pigments were being used for body painting at this early date, before there
is any fossil evidence of modern humans. This could have been symbolic,
or perhaps at this stage it was only to increase the visual impact of the body
as a display, as I discuss in chapter 8.

About sixty pieces of hematite were excavated by Curtis Marean and
his colleagues from cave PP13B on the southern South African coast at
Pinnacle Point, dating from about 160,000 years ago—within the time
range of the first early modern fossils from the other end of the continent, at Herto and Omo Kibish in Ethiopia. The pigments were found
along with possible evidence of composite tools and indications of the
exploitation of marine resources in the form of shellfish—the earliest definitely known. Again there is the possibility that the hematite had a functional rather than symbolic purpose, but consistent selection of the most

brilliant reds suggests that its use was symbolic. Even stronger evidence of symbolic behavior comes from later sites with definite evidence of modern humans. At about 115,000 years, Skhul Cave in Israel has the oldest known symbolic burial, an early modern man interred clasping the lower jaw of a massive wild boar, and this site also has some of the oldest evidence of shell beads, as well as further natural pigments, including ones that have been heated to change either their color or their chemical properties. The 100,000-year-old occupation of Qafzeh Cave near Nazareth also has pierced shells and red ocher, potentially associated with a number of modern human burials, one of which was a child whose body was covered by huge deer antlers. The earliest burials, both modern human and Neanderthal, seem generally to have been of single individuals, although a woman and a child may have been buried together at Qafzeh, while there are claims for a Neanderthal family cemetery at La Ferrassie Cave in France.

The most impressive single site for early evidence of symbolism, however, is Blombos Cave in South Africa, with a record stretching well beyond 70,000 years ago. Blombos is a relatively small cave in sandstone cliffs on South Africa's southern coast, discovered by the archaeologist Chris Henshilwood on his family's land, and excavated by him over the last twenty years. Although many initially doubted the evidence from the site and the claims for its great antiquity, most experts recognize its significance now. As well as a clear stratigraphy, four dating methods, including luminescence applied to quartz grains and heated stone tools, were employed to make this one of the best-dated Middle Stone Age sites in Africa. The stone tools in these levels include Still Bay points, beautifully shaped thin lanceolate spear points, flaked on both sides. They also show the earliest application of a refined stone toolmaking technique known as *pressure flaking*, some 55,000 years before its best-known manifestation in the Solutrean industry of Europe. Slabs of red ocher were excavated from various levels, including the deepest ones, with wavy, fan, or mesh-shaped patterns carefully engraved on them, and the patterns seem purposeful rather than accidental or random. Although the ocher colors faded somewhat after long periods of burial in the cave sediments, experiments show that the engraved lines would originally have been a

vivid bloodred color. Whether this was part of their symbolic importance to the people of Blombos Cave is unknown, but the cognitive
archaeologist David Lewis-Williams compared the patterns with those
made by people today in trance states or under the influence of hallucinogenic drugs.

Alongside the tools and ocher were objects that are easier to interpret
in terms of their symbolic nonpractical significance: beads made from
seashells of the tick (*Nassarius*). Hundreds have now been excavated from
Blombos, and most show signs of piercing, with many of the holes also
displaying signs of wear, where they must have been suspended from a
string or thong. The shells have a natural shiny luster, but the color seems
to have been modified by rubbing with hematite in some cases and by
heating to darken the shells in other cases, so they may have been strung
in different-colored patterns.

The use of tick shells was seemingly widespread across the geographic
range of modern humans at this time, over a distance of more than 5,000
kilometers. I mentioned the shell beads from Skhul and Qafzeh in Israel

An exterior view of Blombos Cave, South Africa.

Looking out to the sea from Blombos Cave.

earlier, and there are at least five sites in Morocco and Algeria where tick shell beads are present in Middle Paleolithic levels, in several cases dated by luminescence or uranium series to between 80,000 and 100,000 years. A different species of *Nassarius* is present on the Mediterranean coasts, but it may be significant that this same genus was selected to make the beads, and in some cases the shells were transported or exchanged over distances of 190 kilometers from the coast. This tradition of using tick shells seemingly continued into the Upper Paleolithic at sites in Lebanon, Turkey, and southern Europe, but there are changes in the style and the materials used, which might reflect how they were being employed socially.

As the archaeologists Francesco d'Errico and Marian Vanhaeran suggested, in the earliest periods with tick shells, and then ostrich eggshell beads, something attractive but readily available—and manufactured without a huge investment of time and skill—was being employed. Although there is evidence that some shells were treated with red ocher and others were heated to darken them, perhaps to create strings of different colors, the perforation in many shell beads was just punched with a stone tool or was even present naturally from when they were collected on the beach. This suggests the beads primarily had a role in reinforcing social networks as items of exchange within groups, perhaps in gift-giving ceremonies, without large differences in social status.

But by 40,000 years ago, in the Later Stone Age of Africa and the Upper Paleolithic of western Eurasia, distinct kinds of beads made of rarer materials began to proliferate, and these often required far more time to manufacture. For example, as the archaeologist Randall White showed, each Aurignacian ivory bead would have taken hours to make, and the skills required were probably not widespread in the group, suggesting specialist artisans. In the case of amber, lignite, fossil, or mother-of-pearl beads, these would have been rare items, traded or transported over long distances across Europe. Even when pierced animal teeth were used as pendants or as grave goods, rarer carnivore rather than the common herbivore species used for food were preferred, and the rarity value at times extended to wearing human teeth as pendants—whether these were from members of the group or taken from enemies as trophies is unknown. By this time many different styles and kinds of beads were in use, even in adjoining regions, suggesting that the beads now served to reflect differential status within the groups (role specialization/wealth/power), and group identity and solidarity in comparison with other groups. During the Aurignacian those other groups could even have included Neanderthals, and there is controversial evidence that the Neanderthals themselves developed pendant and bead use, perhaps for their own social reasons or because of the impact of modern humans, as I will discuss in the next chapter.

For modern humans, if not Neanderthals as well, it seems likely that pigments, beads, art, and music formed part of the signaling used during Paleolithic rituals. Rituals of some kind are present in all known human societies today, and they generally consist of a stylized sequence of events governed by strict rules, where the activity concerned is the focus of attention by the group, whether we are talking about a circumcision ritual, an initiation rite, a wedding, holy communion, an award ceremony, or a funeral. Simple rituals, instinctive or learned, are widespread in the animal kingdom, often to ease tension or encourage social bonding,

Some of the beautifully made Still Bay tools from Blombos Cave.

The most famous engraved ocher plaque from Blombos Cave.

75,000-year-old Middle Stone Age bone tools from Blombos Cave.

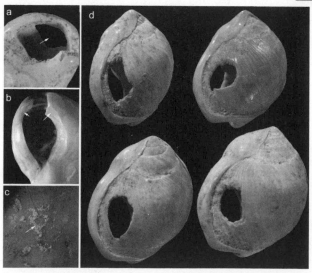

Tick shells that were strung as jewelry from Blombos Cave.

whether it is male baboons gingerly and gently fondling each other's scrotums as a sign of friendship and trust, or a defeated chimpanzee making submissive noises and holding out a hand to the victor (if the gesture is accepted, the dominant chimp will embrace and kiss the supplicant, rather than bite the proffered hand). So humans inherited a ritual base from their primate ancestors, and rituals are so widespread in our species that we can assume they continued to develop for important social reasons. They mark significant events in the life of individuals and of the group, and as groups became ever larger they must have aided symbolic communication and bonding in bigger assemblies, as well as within select subgroups. In order to maximize the impact, and imprint the event on the memory of all concerned, human rituals seem to have become ever more complex and sometimes even risky for the participants, involving deprivation, danger, and fear.

One of the most recent ideas to interpret the cave art of Ice Age Europe combines data from evolutionary psychology and neuropsychology with the evidence of shamanism in contemporary hunter-gatherers. The word *shaman* derives from the Tungus language of Siberia, and these individuals (popularly called *witch doctors*) supposedly possess special powers, giving them access to the spirit world through altered states of consciousness. These altered states may be generated by hallucinogenic plants (for example, in the Shoshone of Wyoming) or by trances induced through pain, deprivation, or (in the San of southern Africa) repetitive rhythmic chanting and dance. Once in the spirit world, shamans often feel they can fly, or travel underground or through water, and they might encounter normal or mythical animals, strange landscapes, ancestors, or gods. They may then transmit messages from the spirit world, foretell the future, or heal the sick. Through their perceived powers, shamans often perform crucial roles in initiation rites and religious ceremonies.

David Lewis-Williams argued that the deep caves of Europe were special places where Cro-Magnon shamans could interact with the spirit world, where the cave walls acted as membranes, and the shape of rocks and crevices signaled the presence of spirit animals or portals between the normal and spirit worlds. The art was thus an expression of altered consciousness, sometimes experienced by groups and at other times by solitary shamans. Just as shamans are crucial members of present-day hunter-gatherer societies, they and their artistic output would have served to represent and even fashion the social and religious systems of the Cro-Magnons. For Lewis-Williams, their artistic imagery reflected the conflicts and hierarchies that were emerging in human societies for the first time. He argued that the roots of such art lie in the fact that *Homo sapiens* may be the only species to remember its dreams (he suggests that Neanderthals probably could not), and in Europe, the presence of the last Neanderthals perhaps triggered cave art as an expression of Cro-Magnon distinctiveness and identity. According to Lewis-Williams, long after the last Neanderthals had gone, the art continued as a deeply entrenched religious system, which reflected the Cro-Magnon societies of which it had become a vital part. Certainly if they were linked to specific religious beliefs, the 25,000-year longevity of the traditions of cave painting and the production of statuettes puts any recent belief systems that we can reliably date far in the shade.

Thus it is likely that some of the deep decorated caves of Europe were used for initiation ceremonies. Perhaps youngsters were even starved or drugged and then taken long distances through a dark cave, to be suddenly confronted with powerful images under torch or lamp light. The experience could have been reinforced by the overwhelming accompaniment of chanting, incense, and drumming; there is evidence that some decorated cave chambers in France and Spain were chosen for their acoustic, as well as their artistic, potential. The even earlier presence of symbolic modern human burials 100,000 years ago implies that burial rituals had already developed by then, and, as we saw, it is possible that the 160,000-year-old Herto skulls were curated, with the child's skull shaped and used as a drinking cup during ceremonies.

All of this indicates that rituals continued to evolve because they enhanced the well-being and survival of individuals and groups. By accumulating the memories of those individuals and the bands to which they belonged, "group memories" could also develop, storing shared information about the tribe and its history. Brain scan studies have shown that areas concerned with working memory (memories not of facts and data but of actions and behaviors) and with behavioral inhibition are both activated during rituals, and the growing importance of rituals to modern humans probably served to reinforce and enhance working memory, mental focus, and the inhibition of "antisocial" (in this context "anti-ritual") actions (that is, ones that would interfere with, or negate the social purpose of, the ritual). By providing a unifying structure beyond individual or subgroup needs, rituals provided a way to direct group behavior, defuse rivalries and tensions, and allow channeled and controlled interactions with potentially hostile neighboring tribes, as long as those tribes understood and abided by the language and etiquette of the ritual concerned—hence replacing suspicion and hostility with trust. Such interactions would have been particularly important for trade, in times of stress (for example, drought), or when sexual partners were needed beyond the tribe.

Once we get to about 40,000 years ago, we can certainly infer the existence of rituals and ceremonies to mark the death of individuals, including multiple burials or special treatment of the dead. Around that date two people were interred separately at Lake Mungo in southeastern Australia: a woman was cremated at high temperature and another adult (sex uncertain) was buried stretched out and with a covering of hematite pigment (perhaps originally on the skin or on some covering material such as hide or bark). Ten thousand years later, far across the inhabited world, the Gravettians started burying their dead with red ocher and elaborate grave goods over an area ranging from Wales in the west (Paviland) to Sungir in Russia. A number of their burials were multiple, and some were extraordinarily rich. As mentioned earlier, at Sungir two children, a boy and girl, were interred head to head, accompanied by hematite, long spears made from heat-treated mammoth ivory, ivory carvings, hundreds of pierced arctic fox canines, and some ten thousand ivory beads that must have been sewn on to their perished fur clothing. The spears probably took weeks to make and the beads many months in total, so these children were highly valued by their group, even in death. And a recent discovery, although not fully published yet, may push the evidence for such behavior back even farther. At a cave in the Tsodilo Hills (Botswana), it's claimed by Sheila Coulson and colleagues that about 70,000 years ago, a six-meter-long rock was shaped to enhance its resemblance to the head of a snake, and the contents of the cave supposedly reflected its long-term use for ceremonies during the Middle Stone Age.

This brings us to the critical question of religion and belief systems, to which rituals are often closely linked. It seems likely that a sense of guilt for social infringements (for example, stealing from a neighbor, hitting a defenseless person who had done no harm) had evolved in early humans, since what appears to be a sense of shame can be programmed into social animals such as dogs and some primates. But only humans have a sense of sin—an infraction not against a person but against a divinely sanctioned law. The law in question may relate to hurting another (for example, adultery or murder) or infringing a religiously enforced code of behavior (for example, combing the hair during a thunderstorm for the Semang peoples of Malaysia or eating pork).

So what could have begun this process of separation from the natural world, and the belief in the supernatural? In *The Descent of Man*, Darwin discussed how his dog barked every time the wind caught a parasol, perhaps because in its confusion it imagined there must be an agent (invisible to the dog) that was causing the movement. Darwin added that such imaginings could have been the source of an early belief in spiritual agencies. Thus the mind-reading abilities we discussed earlier, combined with the human understanding of cause and effect, so essential for activities like toolmaking and hunting, might lie at the root of spiritual beliefs, as argued by both Robin Dunbar and the anatomist Lewis Wolpert. Unexplained phenomena such as lightning, environmental crises, and human illness must have causes, so perhaps invisible spirit forces were at work—as Darwin put it, ones with "the same passions, the same love of vengeance or simplest form of justice, and the same affections they themselves experienced." In particular, once self-awareness had evolved, belief in an afterlife could soon have followed, allowing the mystery of death to be addressed and dealt with—the essence of those who had loved us and looked after us during our lives would surely live on to look after us after they died.

I mentioned shamanism earlier in relation to Lewis-Williams's interpretations of European cave art, and his suggestion that modern humans may be the only species that can remember its dreams, hence providing the imaginative basis for spirit worlds to which humans may gain privileged access. He and others have argued that shamanism is an ancient form of religion, perhaps the very oldest, with an antiquity going back to the African Middle Stone Age at least. In both San and Paleolithic art there are representations of therianthropes (human–animal chimeras—the centaurs of Greek mythology, for example), and in recent depictions these often relate to "soul flights," where the shaman's soul leaves the body during a trance and merges with or is possessed by a spiritually powerful animal. The trances may be brought on by repetitive chanting, dancing, or drumming, by sensory deprivation or sensory overload—for example, through eating, drinking, or smoking hallucinogenic plant compounds. In evolutionary terms, the benefits to the shaman may be obvious—high status and possibly privileged access to group resources or sexual partners—but what are the advantages to the group and the other individuals within it? This brings us to the tricky question of why spiritual beliefs evolved in the first place, and why they seem to have such a hold on humanity, despite occasional and largely unsuccessful attempts to cast them off.

For some, religious beliefs are a pathology—a mass delusion—or they are akin to a virus that perpetuates itself via information imprinted by adults on impressionable young minds. Others argue that spiritual beliefs evolved because they were useful to those who possessed them, and endowed survival on those individuals and their close relatives. Data show that human feelings such as depression, pessimism, and anxiety are handicaps to health and longevity, so religious beliefs that alleviated those "symptoms" could certainly have been favored. Humans do seem to be preprogrammed for religious beliefs, readily taking these onboard, however irrational they may seem to nonbelievers or those of different faiths—and this seems to be as true for adult converts as it is for religiously groomed children. There is disputed evidence that people with strong religious convictions tend to be healthier, live longer, have more surviving children, and are even somewhat wealthier than nonbelievers. If that was true in the past, selection would have favored those with religious beliefs, as long as the benefits outweighed the costs. (Religions or sects that demanded complete sexual abstinence or castration of male followers have understandably not thrived!)

As an example of the social benefits that may have applied in the past, we can return to shamans in tribal societies, who act as spiritual emissaries and appear to have functioned successfully as healers, fortune-tellers, peacemakers, and interlocutors with the world of spirits and ancestors. They may benefit personally through their perceived powers, of course, but they can also act as social enforcers, discouraging aberrant behavior or prophesying in order to lead their groups in new directions. And when we compare recent human societies of all types, there seems to be an association between larger group sizes and the prevalence of morally concerned gods, which could again aid social cohesion and conformism to social norms. Furthermore, modern psychological experiments have shown that religious beliefs can breed selfless behavior (and hence social reputation), discourage "freeloaders," and encourage mutual trust.

From the evidence of burials and symbolic objects, rituals and religious beliefs probably go back more than 100,000 years, but could they actually have been central to the origins of modern humans? A British anthropologist, Chris Knight, certainly thinks so, and in a wide-ranging synthesis of data from present-day anthropology, primatology, and sociobiology, together with archaeology, he and his collaborators argued that women collectively produced a social revolution in Africa over 100,000 years ago. The symbolic use of red ocher began as part of a female response to accumulating social and reproductive stresses caused by the increasing demands of pregnancy, infant and child care, and the need for male provisioning. The bloodred pigment was deployed by menstruating and nonmenstruating women, smeared on their bodies to spread the taboo of menstruation across alliances of female kin. This instituted a "sex strike," which could only be broken when the men returned from collaborative hunts with food to share. Female rituals evolved around the sex strike, male rituals around the hunt (begun under a dark moon, returning at full moon, thus linking menstrual and lunar cycles and the blood of women and of animals), and tribal rituals of celebration and feasting would follow the return of the successful hunters.

I think these ideas are ingenious, and I do believe that human behavior changed in revolutionary ways during the Middle Stone Age, to trigger our expansions within and then outside of Africa. However, I don't think that Chris Knight's views provide the correct explanation or even the correct kind of explanation. This is because I no longer think that there is a single "right" answer to the question of our behavioral origins. What we have seen so far is that there are many interconnected strands to modern human behavior, ranging from our enhanced mind-reading talents, symbolism, and artistic and musical expression to rituals and religion. And, as I will discuss next, we have complex survival mechanisms, which are fueled by our language abilities.

6

Behaving in a Modern Way:
Technology and Lifeways

EIGHT YEARS BEFORE QUARRYMEN CAME ACROSS A STRANGE skeleton in the Neander Valley (Neander thal) in Germany, which gave its name to that whole ancient population, something similar happened in Gibraltar, though the result was rather different. There, the skull of a Neanderthal woman was discovered in 1848, but was then left unrecognized on a museum shelf for the next fifteen years rather than studied and published. So today we talk about Neanderthal man and *Homo neanderthalensis*, rather than Gibraltar Man (or Woman) and "*Homo calpicus*" (a name based on an ancient name for the Rock, suggested by the paleontologist Hugh Falconer in a letter to George Busk in 1863 but never properly published). By then the Gibraltar fossil had made its way to London, and it now resides in a metal safe outside my room. Unfortunately that fossil was blasted from its quarry and no other bones, tools, or associated materials were recovered, though they must have been there. So in 1994 I jumped at the chance of excavating more caves in Gibraltar with a team including Oxford archaeologist Nick Barton and Clive Finlayson from the Gibraltar Museum. Although we never managed to add to the total of two Neanderthal fossils from this tiny pinnacle of limestone, we did find a wealth of evidence of the way of life of these ancient Europeans. It included tools, hearths, food debris, and some of the best evidence yet discovered that the Neanderthals shared a fundamental behavioral feature with us—exploitation of marine resources such as shellfish and seals. That work was published a few years ago and since then even more evidence of the complexity of both Neanderthal and early modern behavior has emerged.

Around 300,000 years ago, the more complex technologies of the Middle Paleolithic began to appear among the descendants of *Homo heidelbergensis* in Africa (*Homo sapiens*) and western Eurasia (Neanderthals). Techniques that required additional steps in tool manufacture became widespread across Africa and western Eurasia, and the first truly composite tools were invented, ones which must have been mounted in or on wooden handles. The wooden handles or shafts almost invariably perished, but traces of adhesive are present on European, western Asian, and African tools. Those used in the African Middle Stone Age were often mixtures of plant gum and red ocher, and the artisans were able to effect changes in their properties through heating and variations in moisture and acidity, implying a high level of knowledge, planning, and thought. Further evidence of these abilities recently emerged from observations of Middle Stone Age tools and from modern experiments.

Archaeologists such as Kyle Brown and Curtis Marean found that they were unable to match the appearance and quality of the many tools they were excavating from the Pinnacle Point Caves in South Africa among the local sources of the silcrete rock from which they had been made. But they finally discovered that their glossier and darker sheen, and finer flaking, only appeared when the tools had been pretreated by being buried under a hearth that was burning for many hours at a high temperature, and then left to cool slowly. Such engineering allowed the removal of longer, shallower flakes and more control of the final shape and cutting edges, and its use on the ancient tools was further demonstrated through physical tests of their fabric, showing they had indeed been subject to prolonged heating. Given the systematic and widespread application of such processes on the Pinnacle Point tools, the results could not have been produced by the tools being left accidentally near a hearth made for other reasons. Not only did this skillful pretreatment lead to tools that looked and performed better, but by improving the quality of the local raw materials, it gave these ancient inhabitants of the southern African coast more options in their choices of rock sources for their tools. This was an essential prerequisite in decisions about where to live, and a sign of their increasing ability to shape their local environment—a key factor in the development of our modern human capacity to adapt to almost any place on Earth.

Fire has, of course, been a vital aid in human survival for at least 800,000 years (based on evidence for hearths at the Israeli handaxe site of Gesher Benot Ya'aqov) and possibly for much longer. As Darwin argued in *The Descent of Man* (1871), "The art of making fire . . . is probably the greatest [discovery], excepting language, ever made by man." It provided warmth and protection from predators, illumination to extend "daylight," and a new social focus as people sat to talk, sleep, or work (and later to sing and dance) around the flickering flames. But the anthropologist Richard Wrangham argued that it had an equally important role in shaping our evolution through the introduction of cooking. In most cases, cooking reduces the time and energy needed to chew and digest foods, although heat also reduces their vitamin content, and nutrients are lost in the fat and water that are driven away. The process not only helped to provide a broader diet and more fuel for a growing and energy-sapping brain, but also reduced the effect of harmful toxins and pathogens such as parasites, bacteria, and viruses that are present in many raw foods. And by adding food to the flames, cooking provided an extra social focus for fire, in that individuals could cook for each other, for partners, kin, friends, and hon- ored guests. Once cooking became central to human life, it would have influenced our evolution, leading to changes in digestion, gut size and function, tooth and jaw size, and the muscles for mastication.

So when did humans first control the use of fire, and when did cook- ing become important? As we discussed earlier, brain size increase and dental reduction had certainly begun in *Homo erectus*, was well devel- oped by the time of *heidelbergensis*, and reached levels comparable to that of modern humans in the Neanderthals. There is disputed evidence for human control of fire dating back to about 1.6 million years in Africa, and stronger support for its presence at about 800,000 years in Israel, and in Britain by about 400,000 years (the site of Beeches Pit in Suffolk). How- ever, the majority of early human sites at this time lack such evidence, which perhaps indicates that the use of fire was not yet ubiquitous among early humans. However, within the last 200,000 years there are many Neanderthal and early modern sites with accumulations of hearths but, interestingly, the associated food debris does not always show strong evidence that the meat was being cooked. For example, at Neanderthal sites I have been involved in excavating in Gibraltar, it seems that the Neanderthals knew about baking mussels in the dying embers of a fire to get them to open up for consumption, but many of the animal remains around their hearths seem to have been butchered and eaten raw.

From fragments of debris preserved in their sites and even around their teeth, we also know that Neanderthals were processing and cooking plant resources like grains and tubers. Similarly, through studies of 100,000-year-old Middle Stone Age tools from the Niassa Rift in Mozambique, Julio Mercader and his colleagues detected the traces of starches from at least a dozen underground and overground plant foods, suggesting that the complex processing of plants, fruits, and tubers, including cooking to remove toxins, was something that had also developed in Africa, providing a vital adaptation as our species traveled around the world. Anna Revedin and her colleagues identified starch grains from wild plants on 30,000-year-old Gravettian grinding stone from sites in Italy, Russia, and the Czech Republic, apparently part of the production of flour, long before the agricultural revolution. The plants included rushes and grasses that, from modern comparisons, were probably exploited at different times of the year, and processed by specialized slicing tools found at the sites. Jiří Svoboda described large underground earth ovens full of hot stones that were in use about 30,000 years ago, at what is now Pavlov in the Czech Republic, to cook huge slabs of mammoth meat. They and surrounding pits, which seem to have been used to boil water with hot stones, were placed within large tents or yurts, to judge by the excavated patterns of holes in the ground. As discussed earlier, such places would have been foci for groups who cooked and ate together.

Just as Chris Knight's model of females bonding had menstruation at its heart, so the anthropologists James O'Connell and Kristen Hawkes argued that the collection and processing of plant resources, especially underground ones like tubers that required specialist knowledge to collect and treat, was critical in catalyzing social change in early humans. Although meat became very important, it was also an unpredictable food resource, so while hunting was left to the men, women—especially those unencumbered with children—developed and shared the skills of gathering and refining plant resources as an insurance policy. So perhaps Darwin's 1871 suggestion that "[man] has discovered the art of making fire, by which hard and stringy roots can be rendered digestible and poisonous roots or herbs innocuous" was actually most appropriate for bands of females. In what has been called the Grandmother Hypothesis, Hawkes and O'Connell proposed that selection would also have favored experienced and postreproductive women who survived for decades after the menopause, something very rare in other primates. These women could have helped to provide for their daughters and other dependent kin, and also act as general helpers, as many grandmothers do today, thus aiding the survival of their genes and the reinforcement of this supportive behavior.

The anthropologist Sarah Blaffer Hrdy took this line of thought farther still with the wider concept of *alloparents*—individuals that regularly took over the provisioning and care of infants and children from the mother. This occurs in other animals, including some primates, and Hrdy believes that the presence of large-brained and dependent infants by the time of *Homo erectus* meant that such supportive social behavior by older siblings and wider kin had already developed out of necessity by then. In her view, such cooperative breeding allowed children to grow up slowly and remain dependent on others for many years, which in turn permitted the evolution of even bigger-brained modern humans. If we compare hunter-gatherer populations today with, say, chimps, there is a huge difference in fecundity: the interval between births averages about seven years in apes but is only three to four years in humans. In apes, mothers do not usually welcome others carrying or even touching their young babies, whereas human mothers are very tolerant of such sharing behavior, eliciting support that Hawkes, O'Connell, and Hrdy argue is the reason humans can cope with such closely spaced births of demanding infants. And Hrdy further suggests that the immersion of human babies in a pool of alloparents would have honed the mind-reading skills and empathy that are so important to our species, faster than anything else. Whether alloparents necessarily included fathers in the past is still unclear, since in the vast majority of mammals, males have little or no specific interaction with their offspring, and the extent of their involvement in infant care also varies widely in humans today. Undoubtedly this would also have depended on the extent of specialized roles in Paleolithic societies; if men were mostly away tracking and hunting, they simply would not have been around much to take on the role of infant care.

Next we will look at the question of the sexual division of labor in early humans and the different views that have emerged.

At one extreme, the archaeologist Lewis Binford provocatively suggested modern humans might have been the first to "invent" the nuclear family, and that Neanderthal social structures could have been more like those of some mammalian carnivores, with packs of males roaming the landscape for meat and living largely separate lives from females (apart from occasional visits to exchange meat for sex). The women were left to bring up their children on what they could gather close to their home bases and nurseries. At the other extreme is the view of the archaeologists Steven Kuhn and Mary Stiner, who argued that hunting big game was a family affair for the Neanderthals, with the women and children joining in, and that by contrast modern humans were the first to develop the patterns of division of labor and distinct roles that we see in hunter-gatherers today. In their opinion, the archaeological record of the Neanderthals showed little evidence of role specialization, and instead the population lived fast, burning energy from a high-calorie diet obtained from hunting large herbivores. Such high-yield food was rich and rewarding, but not always easy to obtain, so the Neanderthals were at the top of their food chain and could only persist in relatively small numbers, at low population densities.

Several lines of argument support such a view. For one thing, from various data the Neanderthals seem to show low levels of sexual dimorphism—that is, males and females were nearly equal in size—which would not be expected if they had very different roles, including male specialization in hunting big game. Second, there are data from chemical analyses of Neanderthal bones (see chapter 3) that suggest they were indeed highly carnivorous, at least in the northern parts of their range. Third, research by Thomas Berger and Erik Trinkaus examined patterns of injury on Neanderthal skeletons and found a high frequency of lesions and fractures, particularly in the head and neck. When they compared the pattern with those in recent and archaeological samples of modern humans, they could not match it, and it was not until they turned to data on injuries in athletes that they could—in rodeo riders, of all people! This did not mean that Neanderthals regularly rode on wild animals, but it did indicate they shared a proximity to hostile wild animals that might bite, butt, kick, roll, or fall on them—and the pattern was found throughout the Neanderthal sample of men, women, and children.

The anthropologist Steve Churchill and the archaeologist John Shea followed the biologist Valerius Geist in arguing that Neanderthals engaged in confrontational hunting, at close range, with wooden thrusting spears—a much more dangerous hunting method than "killing at a distance" with throwing projectiles, arrows, or blowpipes. So if Neanderthal women and children were involved in the hunt, even if only as drivers or beaters, they would have carried a risk of injury from large prey. In contrast, Kuhn and Stiner argue that early modern humans in Africa were able to exist in larger numbers at a greater density than Neanderthals, and in an environment with greater biodiversity. This would have encouraged a varied rather than single approach to procuring food, and the evolution of much more distinct roles for different components of human groups, especially males and females.

As we discussed earlier, meat from large mammals provides rich returns, but it is a supply with inherent risks, both in obtaining it and in relying on such an unpredictable resource. By dividing labor and diversifying food intake to the maximum, modern humans were better able to guarantee where the next meal was coming from, and their reproductive core—women and children—were at reduced risk. Compared with other primate relatives and with what we know of the earliest humans, recent hunter-gatherers have diverse sources of animal protein and fat, of food gathering and processing, and of food storage. Much of this comes from the activities of the old, and from women and children, using snares, nets, and traps to collect small game and tools to extract plant staples. Among recent hunter-gatherers in Australia, Africa, and the Americas, nets have been effective in capturing prey from lizards and small birds up to the size of large deer, and a drive toward the nets is something that almost everyone can collaborate in and enjoy, whether on land or in shallow water. If a surplus of prey is obtained, it can be eaten in ceremonial feasts, traded with neighboring groups, or preserved through drying, smoking, or underground storage.

In the late 1960s the archaeologists Lewis Binford and Kent Flannery both proposed that the "Broad Spectrum Revolution" of recent hunter-gatherers first developed in the last 20,000 years in the Middle East, under the pressure of climate change and increasing population density. In a sense it was forced on the late Paleolithic peoples of western Asia as a way of increasing the carrying capacity of the land from which they lived, and it was seen as a prelude to the domestication of plants and animals that followed soon afterward. But Stiner and Kuhn compared site data covering a much wider range in time and space, and they believe that this ratcheting up of resource exploitation began earlier in human evolution. Archaeological evidence for the broadening of Paleolithic diets in early moderns shows from at least 40,000 years ago, and there is supporting isotopic evidence for this, which I discussed in chapter 3. Grinding tools (sometimes just cobbles) become more common, and, as discussed earlier, they would have been useful for obtaining the maximum benefit (and sometimes also minimum risk of natural toxins) from energy-rich nuts, seeds, and tubers. While large game was still hunted, mainly using projectile spears but later enhanced by atlatls (spear throwers) and bows and arrows, evidence for the exploitation of small game such as tortoises, rabbits, wild fowl, and eggs increased. In addition, food from the sea, the shore, rivers, and lakes became more important.

All of these elements of the diet were already there in some areas and, to some extent, in earlier moderns and in the Neanderthals. (Our work on Neanderthal sites in Gibraltar shows that they were well aware of the value of shellfish, marine mammals, rabbits, nuts, and seeds.) But it seems that for modern humans, such items started to form significant constituents of their diet. And the increasing breadth and processing of plant resources could also have been important in another way. For many hunter-gatherers, starchy foods could have been used to make pastes and gruels for baby food, accelerating the process of weaning, freeing up the mother's time, and allowing alloparents a greater role. In turn, an earlier cessation of breast-feeding potentially returned the mother to the reproductive cycle—a significant factor in the close birth-spacings achieved by modern hunter-gatherers. This, too, could have been a key to the success of the modern human species.

The archaeologist Olga Soffer collaborated in studies of many Czech

Upper Paleolithic sites, and she challenged the prevailing view that these sites contain extensive evidence of the main component of the diet 30,000 years ago: mammoth meat from hunting carried out by Cro-Magnon men. Instead, study of the mammoth bone accumulations suggested that many were probably from animals that had died naturally and were then scavenged some time afterward for their bones and ivory. Moreover, many of those that had been butchered or cooked were either very young or very old individuals, that is, those that were the most vulnerable to natural deaths or other predators, or that would have been the easiest and least dangerous to catch. The implication was that mammoth meat may not have been the main or most reliable item on the menu for a lot of the time. But if not, what supported these large and sophisticated communities in the windswept plains of the last Ice Age? Well, the bones of hares and foxes were common, but microscopic studies of the hearths also revealed residues of plants, fruits, seeds, and roots that were full of starch.

But Soffer also spotted something remarkable impressed on some of the clay fragments that littered the site: delicate parallel lines. The archaeologist Jim Adovasio subjected these to detailed study and found not only many more lines but also crisscross patterns forming a mesh—the telltale traces of woven fibers. Further study revealed the marks of textiles, basketry, nets, cord, and knots. And for those who may be skeptical about interpretations from impressions in clay, fragments of actual flax fibers were discovered in the unusually dry environment of Upper Paleolithic levels at Dzudzuana Cave in Georgia. The work, led by Eliso Kvavadze and Ofer Bar-Yosef, dated some of these fibers as far back as 35,000 years ago. Some had been twisted to make cord or knotted, and others had apparently been dyed in colors ranging from pink to black. The sediments that yielded up the fibers also contained traces of hair and wool from wild ox and goat, as well as the remains of beetles, moths, and mold commonly associated with textiles today.

So the production of twine and other material for sewing clothing and skins, for fastening composite tools together, and for containers, ropes, and netting also seems to have been part of the repertoire of some of the first modern humans in the Caucasus region of western Asia. Such materials would have provided protection from the environment, as well as containers to hold food, and they would have greatly expanded the methods of prey acquisition available to the Cro-Magnons. As we saw, woven nets and traps would also have allowed a wider range of the group to take part in the hunting process, since patience and planning are more important in their use than long-distance travel and physical strength. And such technological changes would also have brought social transformations in the development of specialized roles for the production of things like nets, clothing, and baskets, and the beginning of a whole new range of fashion accessories.

But to return to the Neanderthals and their hunting patterns: despite Kuhn and Stiner's careful arguments, I think that the risks would have been too great for Neanderthal women with young children to have moved far in support of hunting, and, as I explain later in this chapter, there are alternative explanations for the widespread patterns of trauma in their skeletons. Steve Churchill, working with fellow anthropologist Andrew Froehle, expanded on the Neanderthal–modern contrast in subsistence by bringing into the equation climate and the extent of cultural buffering (cultural protection from environmental extremes through heating, clothing, insulated dwellings, et cetera). They suggested that Neanderthals living in Ice Age Europe would typically have needed an extra 250 kilocalories a day compared with a modern human in the same situation, given the higher energetic demands of their lifestyle and to fuel their greater body—and, in particular, muscle—mass. Lower adult energy needs could have given modern humans a breeding and hence a competitive advantage through reduced birth-spacing and greater survivorship compared with the Neanderthals, who were more demanding both on their bodies and on what they needed to extract from their environment. Also, if modern humans were generalists living from a wider range of resources than the more carnivorous Neanderthals, they would have been better able to cope in stressful times.

As we saw, data on injuries in Neanderthals were used by Berger and Trinkaus to suggest that many of these resulted from confrontational hunting, with the further implication that it was not just adult males who suffered in this way. However, they recognized that there were alternative explanations, and I think one in particular needs to be considered, at least as an additional factor: that of interpersonal violence. The pattern of trauma when humans attack each other varies, of course, depending on the weapon used (if not the hands and feet) and on any defense mounted by the recipient. But upper body and head damage invariably predominates, and such injuries may unfortunately be inflicted on women and children as well as men. If a weapon was used, further forensic clues may be left in the form of the weapon itself or traces from it, and we have such data for a couple of Neanderthal wounds, with some interesting speculation surrounding the nature of the assailant.

In chapter 4, I discussed the impact that the discovery of a burial at Saint-Césaire in France had on our views of the Neanderthals in the early 1980s, since it was late in time and associated with the Upper Paleolithic Châtelperronian industry. Recently, Christoph Zollikofer (see chapter 3) and his colleagues studied a scalp injury on the skull, resulting from a slash that was apparently caused by a blade-shaped object. The wound was not deep, although it would certainly have caused blood loss, and it had healed over quite well, suggesting that the individual survived for at least a few months after the incident, perhaps providing evidence for social support among Neanderthals (for more on this, see later in this chapter). Its position suggested that it was not caused by an accident such as a fall or a rock fall, and if the individual was standing upright, it was probably inflicted by a high-energy impact or thrust to the head from the front or back, perhaps by a hafted stone tool such as a spear point.

A second Neanderthal, from Shanidar Cave in Iraq, and known as the Shanidar 3 man, also carries the mark of a spear wound, this time in his rib cage. The partially healed wound was noted by Trinkaus in his study of the skeleton some thirty years ago, but Churchill and his colleagues conducted more detailed studies on the sharp and deep slice in his ninth rib on the left side, including experiments with crossbows that involved firing stone points into pig carcasses. The wound had started to heal, but unlike the Saint-Césaire CSI, it was probably ultimately fatal, either through lung damage or infection, as the spear point may have lodged in the body (although it was apparently not recovered, or at least recognized, during the original excavations). Possible scenarios include a stone knife wound, a hunting injury, or even self-inflicted trauma, but the experiments suggested that the most probable cause was a spear that impacted at a downward angle of about forty-five degrees, most likely one that had been thrown rather than thrust in. Speculating further, Churchill and his team favored the idea that only modern humans had throwing spears with stone tips, and thus they suggested that a modern human rather than a Neanderthal could have been responsible, in an act of interspecies aggression. But could a modern human have been around at the time that the Shanidar 3 man was wounded? That is a major uncertainty since the incident can only be dated to roughly 50,000 years ago, and we cannot reliably place modern humans in Iraq that far back. Equally, it is just possible that the Saint-Césaire individual was confronted by an early Cro-Magnon in France, and these cases can be added to the claimed cannibalism of a Neanderthal child at Les Rois (discussed in chapter 4) as slender evidence that the two species may have had unfriendly encounters.

So we know that the Neanderthals suffered many bodily injuries, and in some cases it seems they must have had social support from others in their group to recover, or at least to prolong their survival. There is a particularly early example of this from the Sima de los Huesos site at Atapuerca in Spain, dating from about 400,000 years ago, where a child with a deformed skull and brain, perhaps caused by an injury sustained before birth, was almost certainly disabled physically and mentally; yet this individual was not rejected at birth and survived the most dependent stages of infancy, dying at around the age of eight, for reasons that may or may not have been connected with the disability. As we discussed earlier, the Atapuerca population lay at the very beginnings of Neanderthal evolution, and it seems that the Neanderthals continued this kind of social support, as may well have been the case for the wounded Saint-Césaire and Shanidar individuals.

Another individual from Shanidar may well demonstrate even higher and longer-lasting levels of social care: the Shanidar 1 man was probably about forty when he died, a very respectable age for a Neanderthal. Yet he had suffered a heavy blow to the left side of his skull and face—perhaps from a rock fall—and as a result may have been partly blind and deaf. Possibly connected with the incident, his right arm had been severely damaged: the upper arm had a badly healed fracture and was withered to a thin stump, and he had completely lost his lower arm and hand. His legs show that he was disabled in walking too, perhaps because the blow to the left side of the brain had caused paralysis on his right side, as may happen in modern injuries of this kind. Despite all those difficulties, he had

apparently survived for many years, implying assistance and provisioning from others. Apes with arm or leg fractures or amputations can
sometimes survive in the wild without social support, but for a Neanderthal living in the Zagros Mountains it seems likely that his injuries
would have been an immediate death sentence without consistent help
from his group.

There are several other examples of survival with impairment in Neanderthals, and also comparable examples from Africa: the 400,000-year-old
Salé cranium from Morocco and the Singa cranium (more than 130,000
years old) from Sudan both show evidence of long-lasting and probably
disabling deformation, yet these individuals survived into adulthood. In
my view, this level of social support probably led to the practice of intentional burial, since, for example, leaving a body on the floor of a cave to
which you might return could entail seeing your father, mother, or siblings
picked over by hyenas or vultures. Later, with repetition and the addition
of ritual, the rise of symbolic burials could have followed, with grave goods
as tributes or offerings to help passage to the spirit world.

To what extent the Neanderthals shared this behavior is still hotly
argued, and a few archaeologists like Robert Gargett even doubt that
Neanderthals buried their dead at all, in which case all the supposed
burials in caves were either accidental or the result of roof falls et cetera.
But I think there is sufficient evidence for some level of ritual behavior in
the later Neanderthals at least, including infants being buried with simple
grave goods. However, it seems likely that one of the most famous examples, which gave rise to the notion that the Neanderthals were the first
"flower people," was the result of other, rather surprising, agencies. After
the Shanidar 4 burial was excavated from this Iraqi cave in 1960, analyses showed that the sediments contained clusters of pollen, suggesting
that bright flowers (perhaps even some with medicinal properties) had
been strewn around the body. However, the zooarchaeologist Richard
Redding subsequently excavated a number of burrows of a gerbil-like
rodent found in the Zagros Mountains near Shanidar and noted that
these animals stored flower heads in their tunnels. In turn, the anthropologist Jeffrey Sommer noticed that the original excavators had reported
rodent bones and burrows around the Neanderthal skeletons; thus it

seems likely that the supposed flower burial of the Shanidar 4 man had a more prosaic and less romantic explanation.

Nevertheless, the care that both Neanderthals and early moderns bestowed on other group members would have had both social and demographic effects, and this may provide further clues about why modern humans were ultimately the most successful of all human species. Earlier, we discussed the distinctiveness of human age profiles compared with those of apes: we have a longer period of infant dependency, reach puberty later, have later ages for first births but closer birth-spacings, postreproductive survival in women is very common, and overall we live longer. This means that humans develop and need much longer-lasting social ties, beyond those of their immediate kin, throughout their lives. There are probably specific genetic bases for our longevity. For example, it has been suggested that unique mutations in a gene for the cholesterol-transporting apolipoprotein E occurred about 250,000 years ago. The variant ApoE3 lowers the risk of many age-related conditions such as coronary disease and Alzheimer's, and it will be interesting to see whether this variant was also present in the Neanderthal genome.

As discussed in chapter 3, the Neanderthals had a human rather than an apelike developmental pattern, but at the same time their lives must have been stressful. About twenty years ago, a nurse turned anthropologist, Mary Ursula Brennan, compared the pattern of growth interruptions in dental enamel formation in Neanderthals and early modern humans and found these indicators of childhood stress were much more common among the Neanderthals. In old age there are further indications of the problems that they and our African ancestors faced (again, from research using teeth—this time to assess the longevity of Neanderthals and early modern humans). While Erik Trinkaus found little difference in survivorship between archaic and modern humans, the anthropologists Rachel Caspari and Sang-Hee Lee came to different conclusions. Their studies were conducted using a technique called *wear seriation*, in which the degree of wear of each molar tooth is used to assess the relative age of an individual. So, for example, the age of third molar (wisdom tooth) eruption was taken to mark adulthood, and when cumulative molar tooth wear indicated an individual was about double that age, they were considered to have reached older adulthood and could potentially have been grandparents. Additionally, Caspari used microCT (see chapter 3) on some dental samples as an aging guide, since the pulp cavities of molars decrease in size through life as dentine is accumulated in them.

Caspari and Lee carried out comparisons ranging from ancient hominins such as australopithecines through to Neanderthals and Cro-Magnons, assessing the ratios of young adults to old adults. They found that only the Cro-Magnons of Europe had a high representation of middle-aged to old individuals (about four times as many, compared with their Neanderthal predecessors in Europe, and even more distinct when compared with earlier humans and prehumans). Interestingly, the Skhul and Qafzeh early moderns were no different from the Neanderthals in their relatively low survivorship to middle and old age. This in turn suggested that cultural, social, or environmental factors—rather than biology— were probably at work in catalyzing the change in age profiles; otherwise the difference should already have been showing in the 100,000-year-old moderns from Israel. If the Cro-Magnons had had more older adults, they would have had more reproductive opportunities, packing extra children into each fertile life span, and there would have been more intergenerational overlap, allowing greater transfer of knowledge and experience down the years. In addition, some data from recent humans suggest that the frontal lobes of the brain, which are closely involved in the planning of behavior, continue their wiring-up until at least twenty-five years of age, so this is something that might only be complete in adults who survived that long. But harking back to the Grandmother Hypothesis and alloparents, these results suggest that their beneficial effects would have barely been felt in early humans, including the Neanderthals. Caspari's study of the seventy-five or so Neanderthals from the site of Krapina in Croatia showed no individuals were likely to have been older than thirty-five at death, so there were not many grandparents around, and that would have been even worse news when so many younger parents were evidently dying before they reached thirty. Thus orphaned Neanderthals would have mainly had to rely on older siblings rather than grandparents for social support.

It was perhaps only with the broadening of food supplies and of those involved in its collection that the change in age profiles could develop in modern humans. And something else of great importance would have been enabled by the overlap of three or four generations in the Cro-Magnons: extended kinship. An example of how important this could have been is shown by the complex kinship systems of many Australian aboriginal groups, which determine not only where individuals are placed in society but what their duties are and how they will be treated. The system determines who can marry whom, what roles they will take in ceremonies, and how they should react to both kin and nonkin (for example, social intimacy, joking relationships, or—cue for many comedians—avoidance relationships such as between a mother-in-law and son-in-law). And when times are hard, groups may need support—or at least tolerance—from each other, such as when a water hole needs to be shared. Then it is critical for negotiators to establish if they are kin or potential enemies by tracking back their genealogies to see if they can find relatives (who may be long dead) in common, or if there is a history of unresolved disputes. All of this requires extensive records and mapping of relationships, which, in the absence of written or digital storage, is only feasible when several generations overlap, in order to provide a kind of collective memory.

In that last example, from Australia, we see the two opposing forces of intergroup relations at work in modern humans—cooperation and conflict—and undoubtedly these have both been important in influencing recent human evolution. I have spent some time discussing the role of mutual social support within groups, but humans undoubtedly also evolved vital mechanisms to defuse potentially aggressive encounters with neighbors. These would have included intermarriage, so that potential enemies could instead become kin, and it is possible that some of the symbolism we see in the Paleolithic—whether it is strings of beads as friendly trade items or cave art intended to signal territorial boundaries—was aimed at managing external relations. The anthropologists Robin Fox and Bernard Chapais developed the argument that the exchange of mates, and in particular the exchange of women, associated with marriage, was the critical evolutionary step in the development of the kinship systems that can be found in hunter-gatherers and pastoralists around the world. Two critical building blocks in such relationships are found in primates: *alliance* and *descent*. Alliance consists of stable breeding bonds, such as a male gorilla and the several females with whom he mates. Descent consists of groups of related individuals, such as female monkeys who share a mother, who bond, and who can acquire the status of their mother and pass it on to their offspring.

But human kinship combines both of these, since the mode of descent (traced through one of the parents) is a mechanism for the construction of alliances. So although offspring disperse, as one sex (usually women) marries outside their immediate group, they maintain their original ties of mutual descent. The change from relatively promiscuous mating to pair-bonding allowed the unique recognition in humans of fatherhood, of paternal relations, and of "in-laws," all of which were essential building blocks in truly human kinship systems. We have little evidence of the kinship systems of the first modern humans or of the Neanderthals (although see chapter 7), but the proliferation of symbolic objects such as beads 80,000 years ago suggests to me that mate exchanges (and most commonly these are exchanges of females) were probably in place between human groups in Africa.

However, the injuries carried by early humans, and especially the Neanderthals, show that encounters with others in the Paleolithic were not always friendly, and although there is less evidence of such wounds in early modern humans, researchers like Raymond Kelly believe that the potential for both conflict and coalitions was also a significant force in the development of modern humanity. I discussed the possibility that only modern humans had projectile weaponry in relation to the rib wound on the Shanidar 3 Neanderthal, and the emergence of "killing at a distance" would have been a threat to humans as much as to hunted prey. Male chimps form aggressive coalitions to carry out lethal raids on other troops, so it is likely that such behavior was part of our evolutionary heritage, and that tools in the form of rocks, clubs, sharp stones, or pointed sticks would soon have been recruited for defense or attack (as in one of the famous opening scenes in Stanley Kubrick's film *2001: A Space Odyssey*). As Darwin put it in 1871: "A tribe including many members who, from possessing in a high degree the spirit of patriotism, fidelity, obedience, courage, and sympathy, were always ready to aid one another, and to sacrifice themselves for the common good, would be victorious over most other tribes; and this would be natural selection." Over the last 130 years, such views have formed the basis of ideas on "group selection" by distinguished researchers ranging from Arthur Keith and Raymond Dart to Richard Alexander and James Moore.

But from the 1970s onward, work by biologists such as William Hamilton, Robert Trivers, and Richard Dawkins emphasized the selfishness of genes and undermined the basis of many previous formulations of group selection. Selection acts only on genes or individuals, not populations, and while altruism (selflessness) can evolve, it will only be favored in genetically closely related groups. Mathematical tests showed that group selection would fail when there was even a small amount of migration between groups, or when "cheaters" exploited the benevolence of others to propagate their own genes. However, more recently, biologists and anthropologists such as Paul Bingham and Samuel Bowles have returned to the issue by recruiting weaponry and genes to the cause of group selection. The argument goes that by coming together to use effective projectile weaponry, individuals reduced their separate risks, and thus coalitions of warriors would have been advantageous for group defense and offense. Bingham proposed that this development would also have been important within societies by deterring free riders who tried to reap the rewards of group membership without contributing their fair share of commitment to the associated costs or risks. However strong individually, they could soon be brought into line when faced with a coalition of spear-armed peers, who could act as general enforcers of within-group rules and solidarity.

Bowles posited the idea that if Paleolithic groups were relatively inbred and genetically distinct from each other, and warfare between groups was prevalent, then group selection through collaborative defense and attack could evolve and be maintained. Without warfare, a gene with a self-sacrificial cost of only 3 percent would disappear in a few millennia, but with warfare, Bowles's model showed that even levels of self-sacrifice of up to 13 percent could be sustained. He used archaeological data (although mainly post-Paleolithic) to argue that lethal warfare was indeed widespread in prehistory, and that altruistic group-beneficial behaviors that damaged the survival chances of individuals but improved the group's chances of winning a conflict could emerge and even thrive by group selection. Moreover, the model could work whether the behavior in question was genetically based or was a cultural trait such as a shared belief system. As mentioned above, Bowles's archaeological data do not come from the Paleolithic, but there is one observation that does resonate with his views: the French archaeologist Nicolas Teyssandier noted that the period of overlap of the last Neanderthals and first moderns in Europe was characterized by a profusion of different styles of stone points. This might reflect a sort of arms race to perfect the tips of spears, perhaps to hunt more efficiently, but equally this could suggest heightened intergroup conflict.

Social relations, cooperation and conflict, food acquisition, and changing age profiles could all have been important in shaping modern humanity, but one of the markers of *Homo sapiens*—language—was undoubtedly a key factor. For the primatologist Jane Goodall, the lack of sophisticated spoken language was what most differentiated the chimps she studied from us. Once humans possessed this faculty, "they could discuss events that had happened in the past and make complex contingency plans for both the near and the distant future . . . The interaction of mind with mind broadened ideas and sharpened concepts." Despite the rich repertoire of communication in chimps, without a humanlike language "they are trapped within themselves."

So how could such a critical thing as language evolve in humans, and was its evolution gradual or punctuational? Darwin certainly favored a gradual evolution, under the effects of both natural and sexual selection. He wrote in 1871:

> With respect to the origin of articulate language . . . I cannot doubt that language owes its origin to the imitation and modification of various natural sounds, the voices of other animals, and man's own instinctive cries . . . may not some unusually wise ape-like animal have imitated the growl of a beast of prey, and thus told his fellow-monkeys the nature of the expected danger? This would have been a first step in the formation of a language.
>
> As the voice was used more and more, the vocal organs would have been strengthened and perfected through the principle of the inherited effects of use; and this would have reacted on the power of speech. But the relation between the continued use of language and the development of the brain has no doubt been far more important. The mental powers

in some early progenitor of man must have been more highly developed than in any existing ape, before even the most imperfect form of speech could have come into use, but we may confidently believe that the continued use and advancement of this power would have reacted on the mind itself, by enabling and encouraging it to carry on long trains of thought. A complex train of thought can no more be carried on without the aid of words, whether spoken or silent, than a long calculation without the use of figures or algebra.

In contrast to Darwin's gradualist evolutionary views, the linguist Noam Chomsky has long argued that modern human language did not evolve through Darwinian selection; in a sense, for him, it is an all-or-nothing faculty, emanating from a specific language domain in the brain that may have appeared through a fortuitous genetic mutation. He believes that all human languages, however different they may sound at first, are structured around a universal grammar that is already present in the brain of infants and which they use intuitively to interpret and then re-create the patterns of speech presented to them by the group into which they are born. The evolutionary psychologist Steven Pinker has shared some of Chomsky's views, in particular that there is a specific hard-wired domain for language capabilities in the brain. In his opinion, this domain generates *mentalese* (a term created by the cognitive scientist Jerry Fodor), an underlying and innate mental code out of which all human languages can be forged. However, Pinker parted from Chomsky in arguing that gradual genetically based change (comparable to that which eventually led to complex eyes) could have evolved the human "language organ" and its language-generating systems, in a series of evolutionary steps, with selection (either natural or sexual/cultural) favoring increased richness of expression.

Earlier I discussed the view of the archaeologist Richard Klein that there was a punctuational origin for modern human behavior in Africa about 50,000 years ago, and, to an extent, his views can be compared with Chomsky's. Klein critically assessed the evidence for "modern" behavior prior to 50,000 years and found it unconvincing. In his view it is only after that date that a consistent pattern of finds demonstrates the presence of things like increasing diversity and specialization in tools, undoubted art, symbolism, and ritual, expansion into more challenging environments, diversification of food resources, and relatively high population densities. As a trigger, he suggests there may have been "a fortuitous mutation that promoted the fully modern brain ... the postulated genetic change at 50 ka fostered the uniquely modern ability to adapt to a wide range of natural and social circumstances with little or no physiological change." He further speculates that this brain rewiring may have rapidly facilitated the full language capabilities of *Homo sapiens*, which up to then had been little different from those of earlier humans, and, as he recognizes, this is something that is very difficult to demonstrate from the fossil and archaeological record. Although I disagree with Klein about a unique "switch" that turned on modern human behavior, I agree with his views on the critical importance of language to our species.

However, there could have been premodern languages in earlier humans and in the Neanderthals. Robin Dunbar and the anthropologist Leslie Aiello argued that human language perhaps first developed through "gossip," as a supplement to (and eventually a replacement for) social grooming. The activity of fur grooming is performed on a one-to-one basis by many primates to help maintain their relationships and social cohesion. Dunbar and Aiello speculated that without the benefit of language, the burgeoning size of *Homo erectus* groups would have required individuals to expend up to half their time on individual social grooming, leaving little time for other vital activities. But by allowing groups of early humans to chatter to each other, a primitive language could have facilitated social intimacy and cohesion, freeing up time otherwise spent in grooming.

In contrast, the psychologist Michael Corballis returned to Darwin's views on the importance of gestures as precursors of language, arguing that the brain areas that are important for language production in humans are actually concerned with manual actions in other primates. Similarly, the psychologist Michael Tomasello sees language as having evolved as a practical social tool for communication regarding requests, information, and cooperation. In his view, speech is the ultimate, and perhaps ultimately evolved, component of what we call human language, but it could have been preceded by gestures, just as it often is in human babies. Indeed, there is considerable evidence that we do communicate with each other—sometimes unconsciously—through our body language and posture, still an important vestige of our prelinguistic primate heritage. Other researchers have argued for a link between brain coding for toolmaking and for language, since both are sequential processes involving intention and finely controlled muscle actions; thus children learn to manipulate and assemble words as they are also manipulating and assembling objects. And it is certainly possible that parts of the brain that were already there and fulfilling different functions were co-opted to cope with the growing demands generated by language for storage, processing, and muscle control.

From my perspective, modern human language probably evolved out of growing social complexity over the last 250,000 years to bolster mind reading and communication, and I agree with the archaeologist Steven Mithen that by enhancing cognitive fluidity, language took modern humans into new and shared worlds that were unknown to our ancestors. The Neanderthals must have been highly knowledgeable about the world in which they lived, too (for example, about the materials from which they made tools and the animals they hunted). But in my view their domain was largely of the here and now, and they did not regularly inhabit the virtual worlds of the past, the future, and the spirits. After our evolutionary separation about 400,000 years ago, we and the Neanderthals traveled down parallel paths of developing social complexity and, with it, developing language complexity. For whatever reasons, we traveled farther, and the Neanderthals came to the end of their long road about 30,000 years ago.

Attempts have been made by scientists like Philip Lieberman and Jeff Laitman to reconstruct the speech capabilities of Neanderthals and other early hominins, based on the shape of the base of their skulls and the position of critical anatomical landmarks. In their view, modern humans possess a uniquely shaped tongue, throat, and vocal tract, facilitating the range and complexities of sounds required for fully human speech. Darwinian selection must have operated on variations in skull form, leading to its restructuring, and the modification of the throat from its previous dominant functions of breathing and swallowing toward the complexities of modern human speech capabilities, as long as the newly acquired vocal skills conferred advantage. The price we paid for this reshaping of priorities in the throat is a greatly increased risk of choking on our food

compared with chimps and early hominins. In the case of the Neanderthals, their vocal tracts and capabilities were apparently closer to those of
a two-year-old child than a modern adult; however, there is no doubt
that if the Neanderthal brain was coding for complex language, even
that kind of vocal tract would have done the job, albeit with a more
restricted repertoire of sounds.

Returning to the checklist of modern human attributes that I discussed in the previous chapter, in my view modern humans had developed most of them by 60,000 years ago, even if they were not always
present as a package all of the time from the evidence we have found so
far: that is, complex tools, the styles of which may change rapidly through
time and space; long-distance transport of valued materials such as
stone, shells, beads, amber; evidence for ceremonies or rituals from art,
structures, or complex treatment of the dead (the latter inferred from the
Skhul and Qafzeh symbolic burials). But for some of the attributes, the
evidence so far is either equivocal or only partly there, including formal
artifacts shaped from bone, ivory, antler, shell, and similar materials;
greater complexity of food gathering and processing procedures, such as
the use of nets, traps, fishing gear, and complex cooking; art, including
abstract and figurative symbols; and structures such as tents or huts for
living or working that are organized for different activities such as toolmaking, food preparation, and sleeping and for hearths. Regarding higher
population densities approaching those of modern hunter-gatherers, I
will discuss this aspect later, in chapter 7, from the perspective of genetic
data. And regarding increased cultural "buffering" to adapt to more
extreme environments such as deserts or cold steppes, it may be that this
aspect evolved more gradually as modern humans grew in numbers and
dispersed to the ends of the Earth.

In chapter 4 I discussed how fossil foot bones from Europe and China
showed that early moderns in both regions had seemingly discovered the
benefits of footwear, and patterns of objects, such as sewn beads, pins,
and toggles, in Cro-Magnon burials imply the existence of fitted clothing, as does the presence of eyed bone needles. Clothing would have been
of great value to humans in colder climates, and although direct evidence for it has perished, it seems likely that the Neanderthals had the

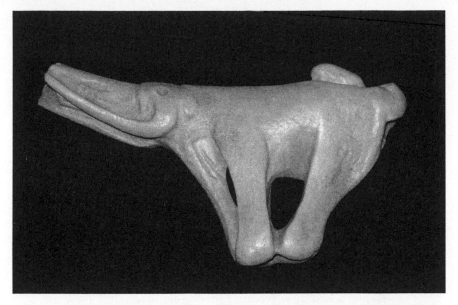

Replica of a mammoth carved from reindeer antler, from the Montastruc rock
shelter, France. The original formed part of a spear thrower (atlatl)
and is about 14,000 years old.

skin-working technology required to make at least basic clothes to keep
the cold and wet off their bodies.

Many modern peoples in tropical and subtropical regions often wear
little or no clothing, beyond what may be required for reasons of mod-
esty or tradition, and humans do have the ability to adapt physically to
colder climates. When Darwin and the *Beagle* visited the bitterly cold
regions of Tierra de Fuego at the subpolar tip of South America, he was
astonished to see that the native peoples wore little or no clothing and
slept naked in the open. Native Australians also have physical adaptations that help them sleep at night in the outback, but interestingly Europeans seem physiologically poorly adapted to the cold—something they
do not seem to have acquired from the Neanderthals, despite the likelihood of interbreeding. However, cold conditions were also present in
Africa, in the highlands and in cloudless areas at night, and basic clothing and warm bedding would have been advantageous at times.

With the lowered temperatures of the last Ice Age, moderns in parts of Africa 60,000 years ago would certainly have benefited from clothing and warm bedding. Although we have no direct evidence of these from that time period, we do have some genetic clues. Humans are infected by head and body lice, and while they both feed by blood-sucking through our skin, the latter live and lay their eggs on clothing and bedding, a fact that both Mark Stoneking and Melissa Toups and their colleagues utilized for evolutionary studies. Both teams reasoned that the origin of the distinct form of body lice probably corresponded to the opportunity provided by the regular use of clothing and bedding. Using mtDNA molecular clocks of lice evolution in humans and apes, they estimated the origin of body lice to between 80,000 and 170,000 years ago. This in turn suggested that bedding and clothing could already have been in use by modern humans in Africa, and they then took this valuable innovation with them when they left their ancestral homeland—together with the lice. However, there is another possibility to consider. The archaeologist Timothy Taylor quite rightly highlighted the invention of the baby sling as a crucial development for the way that it liberated women from the restrictions of baby carrying and the confines of static nursing, and he considers this must have happened relatively early in human evolution. But if the innovation actually happened with modern humans, in Africa, this could have provided a new home and jumping-off point for those pesky lice.

A more welcome fellow traveler on the modern human diaspora from Africa may have been the dog, the first known domestic animal. There is evidence that Aurignacian people living in Goyet Cave, Belgium, already had large dogs accompanying them about 35,000 years ago. The dogs were anatomically distinct from wolves in their shorter and broader snout and dental proportions, and isotope data suggest that they, like the humans, were feeding off horses and wild cattle. Moreover, ancient dog DNA was obtained, which showed that the Belgian dogs were already genetically diverse and that their mitochondrial sequences could not be matched among the large databases of contemporary wolf and dog DNA. These findings are important because they suggest that dog domestication had already been under way well before 35,000 years ago.

So where were the first dogs domesticated? That is a very difficult question to answer from modern dog and wolf DNA, with no clear single link between the two species or subspecies. Indeed, it is quite likely that domestication happened more than once, in different regions and from distinct wolf stocks. Given the Pleistocene range of the wolf, the first event might have occurred in western Asia, soon after modern humans reached there about 55,000 years ago or, as some DNA data suggest, farther east in Asia. It may have happened through the adoption of cubs, or perhaps through a gradual relationship of tolerance as wolves hung around human campsites. With selective dog breeding and adaptation to each other, a special relationship probably developed quickly. (In Russia, silver foxes were bred to behave like dogs over a period of only fifty years.) Ancient dogs obviously bred successfully, and for humans the advantages of having them to provide extra (and more acute) eyes, noses, and ears, as well as speed and fangs, were clear. And coevolution occurred—dog brains are on average some 25 percent smaller than wolf brains, and yet they have many skills that wolves lack. For example, even as puppies, they can follow human pointing to find hidden objects, and their powers of attention and imitation match or exceed those of apes, suggesting quite sophisticated cognition and mind reading in the social domain. And if Aurignacian Cro-Magnons were regularly accompanied by dogs, could this have been another advantage they had over the last of the Neanderthals?

There are some other important issues in the Out of Africa dispersal related to behavior that also need to be addressed. If and when modern humans encountered the Neanderthals, how much would behavioral differences between them have affected the way they saw each other? Would they have perceived each other simply as other people, enemies, or even the next meal? We don't know the answer, and it may have varied from one time and place to another, especially given the vagaries of human behavior. These populations had been diverging from each other for much longer than any modern human groups who encountered each other in the Americas and Australia during the colonial "Age of Discovery." In my view there were probably deep differences in appearance, expression, body language, general behavior, and perhaps even things like smell, which would have impinged on how the Neanderthals and early moderns perceived each other. As an example that seems to apply even to the relatively closely related populations of Europe and the Far East today, there is some evidence that these groups read facial expressions somewhat differently—Europeans using the whole face, and East

Asians focusing more on the eyes as cues for ascertaining mood. As a result, Asians (in an admittedly small sample) were more likely to read European facial signals for fear as indicating surprise, while they more often took disgust to indicate anger. If such differences have arisen within modern humans in the last 50,000 years, the potential for misreading between Neanderthals and moderns might have been even greater. Such factors, as well as possible differences in language, symbolic communication, and social structure, would have been every bit as important as the physical ones in determining whether interbreeding happened, and what were the fates of any offspring that may have resulted from it.

What the symbolic repertoire of Neanderthals actually was remains a subject of hot debate. While early moderns in Africa seemed to have preferred the bloodred signaling of hematite, there is evidence (discussed shortly) that groups of Neanderthals in Europe utilized dark pigments such as manganese dioxide and even pyrite. Some of this may have been for functional reasons, as has also been suggested for hematite, such as treating hides and mixing with resins to form an adhesive. But with emerging genetic data that some Neanderthals were pale-skinned (see chapter 7), it might well be that red pigments showed up best on dark-skinned Africans, while black pigments would have been favored by lighter-hued Neanderthals. Of course, even marking skin with pigment could have had a functional purpose in the Neanderthals, camouflaging them for ambush hunting, but there are other controversial indications that they were signaling symbolically—but to whom?

The Grotte du Renne (Reindeer Cave) at Arcy in France is a site that, like Saint-Césaire, has demonstrated the association of Neanderthals with the "advanced" Châtelperronian industry. There are not only the characteristic stone tools of this industry but also parts of a Neanderthal child's skull and isolated Neanderthal teeth. But more surprisingly there are animal teeth pierced to make pendants, and fragments of worked bone and mammoth ivory. These latter items are, of course, also characteristic of the Aurignacian industry, which seems to have been the product of early Cro-Magnons. So what was going on here? There are several possibilities, each of which has support from one group of archaeologists or another. One is that the symbolic objects were not made or used by Neanderthals but were in fact the products of modern humans—either ones who briefly visited the site and left items like the pendants there or ones who lived there later, following which the cave deposits became mixed or were not excavated well enough to distinguish the separate occupations. In support of this view is recent radiocarbon dating of the site that suggests that the Châtelperronian deposits in particular suffered disturbance and mixing. But against it, there is the fact that other Châtelperronian sites in France and Spain show similarly "advanced" stone and bone tools like those at Arcy, and at least one other contains pendants.

A second point of view is that the symbolic items were made by moderns but had been traded into Neanderthal groups, or that the Neanderthals had picked them up in a nearby Cro-Magnon site. A third suggestion is that the Neanderthals were acculturated by contemporaneous moderns— that is, they were influenced by them and were, for example, copying their jewelry styles. And a fourth idea is that the Neanderthals were in fact continuing their own independent tradition of developing complexity, and undergoing a parallel process of becoming "modern"—but in a Neanderthal way!

Neanderthals used pigments too—these are of manganese dioxide
from Pech-de-l'Azé in France.

In support of that last viewpoint, there is arguably even stronger evidence of complex Neanderthal social behavior from two Middle Paleolithic cave sites in southeastern Spain (Cueva Antón and Cueva de los Aviones). There, museum studies and excavations led by the archaeologist João Zilhão found seashells that had apparently been used symbolically. Cockle and scallop shells with natural holes in the right places to be strung as pendants had been collected and transported inland. Some of the shells had light or dark pigments stored in or painted on them, and a thorny oyster shell contained ground pigment that had apparently been mixed with pyrite as a glittering cosmetic. While the painted scallop from Cueva Antón was probably less than 40,000 years old, and therefore might reflect Cro-Magnon influence, the Cueva de los Aviones material dated to about 50,000 years ago, seemingly too old for that explanation to apply. But in either case these sites represent strong evidence that at least some Neanderthals were expressing themselves symbolically, seemingly as much as many Middle Stone Age Africans, and I will return to this question in the next chapter.

In 1993 Clive Gamble and I argued that the Neanderthals had absorbed aspects of Cro-Magnon culture, but while they could "emulate . . . they could not fully understand." Now I would say instead that if the Neanderthals were making or just using objects like pendants, they were participating in symbolism just as the moderns were, whether they were signaling within their own groups or to others, who might at times have even included early Cro-Magnons. And if that is so, the idea that these aspects of modern human behavior resulted solely from genetic changes in African Middle Stone Age peoples must be wrong, unless the Neanderthals had undergone similar mutational changes in their own evolution, or they had acquired the modernizing genes by hybridization—a subject I will discuss in more detail in the next chapter.

7

Genes and DNA

LIKE MANY PEOPLE I AM CURIOUS ABOUT MY ORIGINS AND SO WAS pleased when the geneticists Bryan Sykes and Alan Cooper wanted to sample and determine my DNA—or at least the tiny bit of it contained in the mitochondria of my cells (mitochondrial DNA). But they had a practical purpose in mind as they were pioneers in the extraction of DNA from human fossils, and they wanted to be able to exclude any of my DNA that might be contaminating the fossils I had handled, or even briefly touched. They were fortunate as I had a couple of unusual mutations in my mtDNA, which makes it very recognizable, but it was still somewhat shocking to find that my DNA had left a contaminating trail across the museums of Europe! As Alan Cooper is jokily fond of accusing paleoanthropologists, in terms of the contamination of fossils he has tried to study, "You are all very dirty people!"

In this chapter we will look at the huge amount of genetic data about the evolution of our species and our diversity now being generated, and address the origin and significance of regional ("racial") differences. The genetic data can be used to look at the demography of ancient humans in Africa, the size of our ancestral pool of people, and the numbers that may have left Africa to found the populations of the rest of the world. They can also be used to estimate dates for events in our evolutionary history, such as our split from the Neanderthals and when our modern human ancestors first moved out of Africa. In addition, in the last decade, scientific breakthroughs gave us tiny but invaluable glimpses of the genetic makeup of the Neanderthals and are now providing a nearly complete Neanderthal genome to compare with ours and with that of chimpanzees. This three-way comparison will illuminate what makes each species really distinct and will lead the way to reconstructing, at least to some extent, what the Neanderthals looked like in the flesh, and perhaps even the humanness of their brains and ways of thinking. Along with discussion of the genetic data, I will give my views about the evidence of mating between Neanderthals and modern humans.

Charles Darwin and his contemporaries had no real knowledge of the mechanisms behind the inheritance of bodily characters, and their predominant ideas were of blending traits between the two parents—and in Darwin's case, that each cell in the body gave out *gemmules*, which agglomerated to reconstitute individuals of similar structure in the next generation. As is well known, while Darwin was writing on such matters, the monk and scientist Gregor Mendel was conducting experiments on heredity in Brno (Czech Republic), using peas and bees. He realized that much of inheritance was particulate rather than blended, and that characteristics (often in several alternative states) were inherited following certain rules. Mendel's work was largely overlooked for another thirty-five years but was rediscovered around 1900, sixteen years after his death, by which time the units of inheritance were known as *genes*.

Half a century after that recognition, the structure and role of deoxyribonucleic acid (DNA) in the makeup of genes was discovered, and the modern science of genetics began to take off. It was realized that the ability of DNA to replicate itself resided in its unique twisted ladder of paired bases. The chemical base adenine (A) was always paired with thymine (T) across the strands of DNA, and cytosine (C) with guanine (G). Thus when the ladder splits for replication, each half can form a template from which its whole structure can be re-created. DNA research has become increasingly important for anthropology in studies of the evolution of primates and their present social structures, and for humans in terms of our affinities to the other primates, to population relationships today, and to reconstructions of our evolutionary history.

Nowadays, our close kinship to the African apes is well established—something that would undoubtedly have pleased Darwin and his close ally Thomas Henry Huxley. But before the impact of genetic studies, it was normal practice for anthropologists to argue that although we were undoubtedly related to the great apes biologically, our special "human" features, such as walking upright, having a large brain, making tools, and speaking, fundamentally set us aside from them. This meant that we were justified in classifying humans as a separate zoological family (the hominids) as distinct from the apes (pongids). Moreover, it was believed that our special features must have taken a very long time to evolve, so many anthropologists favored the idea that our lineage split from that of the apes more than 15 million years ago.

This view has been swept aside in the last twenty years by a wealth of genetic data that suggest that the chimpanzees (common and bonobo) only differ from us in about 2 percent of their genetic material. The actual figure given varies because experts differ in the way they count up the data; for example, whether considering total DNA sequences, including regions of DNA that do not seem to be functional and can be duplicated many times over, or by comparing sequences that can be shown to be precise equivalents to each other, or by restricting the calculation to "functional" or coding DNA regions. Regardless, the differences are comparable to those found between closely related mammals such as African and Indian elephants, horse and zebra, or jackal and wolf. Such similarity implies that there must be a close evolutionary relationship, and calibration (age estimates) using the fossil record and the genetic distances involved suggests that our line of evolution and that of the chimpanzees may only have separated about 6 million years ago. This view began to gain ascendancy about thirty years ago, following the pioneering work of Allan Wilson and Vince Sarich, who conducted studies using genetic differences in the protein albumin, showing that the Asian orangutan was less closely related to us than were the African great apes. This close relationship is now often recognized by admitting chimps (and less consistently the gorilla as well) into the hominid family, along with us and our immediate extinct relatives.

Large-scale comparative sweeps of the genomes of humans and chimpanzees show that the vast majority of the 3 billion or so "letters" of our genetic codes are shared, but the rare stretches of distinct DNA are beginning to yield information of great evolutionary interest. Some are clearly related to the various past epidemics to which we and our ape relatives have been exposed, for example, in conferring resistance to retroviruses like HIV, but others can be related to physical changes. For example, a group of 118 bases known as *human accelerated region 1* (*HAR1*) is virtually identical in animals as different as chickens and chimps, with only two coding differences, but humans have accumulated eighteen further mutations. Experiments showed that this DNA sequence is important in building the structure and connections of the cerebral cortex, the wrinkled outermost layer of the brain that is so important for human intelligence (see chapter 8). Many other genes involved in the growth of the brain as a whole, such as ASPM, CDK5RAP2, CENPJ, and MCPH1 (microcephalin), also show accelerated change compared with chimps, and we will return to the last of these—microcephalin—shortly.

Interestingly, many of the DNA sequence differences that are accentuated between us and our closest living relatives are not concerned with direct changes in, say, the structure of a protein or enzyme. Instead, insertions of what are called *transposable elements* affect portions of the genetic code by acting as switches in turning functional genes on and off. If the direct products of the DNA can be compared with the ingredients in a recipe, these equally important switches in regulatory genes can be seen as altering the instructions for exactly how the meal is to be cooked, which will produce different results (for example, chimps or humans) from a similar recipe (our DNA coding). Thus the *human accelerated region 2* (*HAR2* or *HACNS1*) drives gene activity in building the structure of the wrist and hand bones before birth, and it is likely that these novel DNA changes in humans contribute to our distinctive hands and their greater dexterity, compared with those of chimps and gorillas.

As well as comparing our DNA with that of our closest living relatives, the chimpanzees, we can infer an increasing amount of information about our evolutionary past from the DNA of living humans, since each of us carries an ancestral record locked up in our genes, something much more detailed than a set of parish records, and one that goes much farther back in time. Because DNA is repeatedly copied, especially when it is passed on from parents to their children, copying mistakes are made, and if the changes are not greatly disadvantageous or lethal, these mutations are then also copied. Thus they can accumulate through time and allow us to follow particular lines of genetic evolution, and to estimate the time involved in their accumulation.

For our purposes, there are three kinds of DNA that can be studied. The first type is called *autosomal DNA*. This DNA makes up the chromosomes contained within the nucleus of our body cells, but excluding the special case of the male-related Y-chromosome, which we will come to later in this chapter. It contains the blueprints for most of our bodily structures, and we inherit a combination of it, with our parents making contributions of about 50 percent each. Autosomal DNA also contains many long segments of so-called junk DNA, which do not code for features such as eye color or blood group type. These segments nevertheless get copied, along with the coding DNA, and mutate through time too. Despite their "junk" nickname, some are known to operate as genetic switches, and they can give us valuable information on evolutionary relationships. In fact these sequences are generally more useful in evolutionary and population studies because they may not have been so affected by the distorting consequences of selection, which is strongest on functional DNA—that is, containing genetic code (although junk DNA can be affected when it is structurally linked to functional DNA that is under selection).

The second type of DNA lies on the *Y-chromosome,* which determines the male sex in humans. Normal females have twenty-three pairs of chromosomes, including a pair of X-chromosomes, whereas normal males only have twenty-two pairs, plus an X-chromosome (inherited from the mother) and a Y-chromosome (inherited from the father). The DNA on this chromosome can be used to study evolutionary histories in males only, without the complication of inheritance from two parents that comes with the study of autosomal DNA—rather like the continuity given by male surnames in many societies.

The third type of DNA is the now-famous *mitochondrial DNA* (*mtDNA*), which is found outside the nucleus of cells and which is inherited through females only. Although this last type of DNA has attracted the most attention in the media and popular science—because it gives such a clear signal of ancestry—analysis of the more extensive autosomal DNA and its products (most of our bodies' vital constituents such as organs, proteins, enzymes, antigens) has a much longer history in evolutionary studies. For example, a study of ape and human blood proteins led to the first suggestion of a later divergence between humans and African apes, compared to Asian apes.

As its name suggests, mitochondrial DNA is found in the mitochondria. These little bodies are the power stations of the cells, turning nutrients into usable energy for the cells to do their work. Their DNA is passed on in the egg of the mother when it becomes the first cell of her child, and little or no DNA from the father's sperm seems to be incorporated at fertilization. This means that mtDNA essentially tracks evolution through females only (mothers to daughters), since a son's mtDNA will not be passed on to his children. The molecule of mtDNA is shaped in a loop and consists of about 16,000 base pairs. Only some of these are functional—that is, contain genetic code to produce specific proteins such as cytochrome—and the rest of the mtDNA is therefore much more prone to mutation. Thus mtDNA generally changes at a much faster rate than nuclear DNA, making it ideal for studying recent events and short-term evolution. As mentioned in the introduction, prior to the recovery of Neanderthal DNA, the biggest single impact of genetic data on research on human evolution came in 1987, with the publication of Cann, Stoneking, and Wilson's study of mtDNA variation in modern humans. I described how the work came under heavy attack, especially from disgruntled multiregionalists, but the increasingly detailed analyses carried out since then have shown that the 1987 conclusions were essentially correct, even if they were somewhat overinterpreted.

Some calculations now place the last common mtDNA ancestor (Eve) at less than 150,000 years old, and it is clear that across the whole human species today, our mtDNA varies far less than is the case in great ape species. This has led to the idea that a recent bottleneck—a drastic drop in population—pruned the variation previously found in the modern human line. However, in line with the alternative nickname for Eve—"lucky mother"—some geneticists have explained that this pattern could have occurred purely by chance if just one woman from those ancient times was lucky enough to have a fertile chain of female offspring through to the present. Thus all other mothers from that time ended up unluckily (in terms of the continuity of their mtDNA) having no surviving children, or just boys, or daughters who failed to provide the necessary

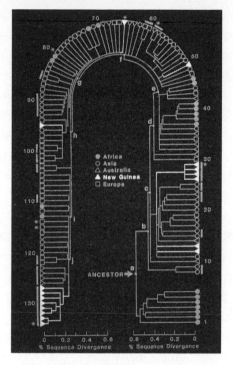

The famous mtDNA tree published
in 1987.

ongoing chain of fecund females. From that perspective Eve was not a
special female and did not necessarily live in special times, but she gained
her unique status retrospectively, through her mtDNA's good fortune.
And we should remember that while this female gave rise to all mtDNA
variants known in humans today, many other individuals have contributed their Y and autosomal DNA to succeeding generations. MtDNA is
important because it provides such a clear signal of ancestry and descent,
but it is effectively inherited as a unit like a single gene, and all our gene
variants have their own history, converging back (coalescing) to ancestral genes at various times in the past. Some of those genes have developed very recently, within smaller or larger segments of the modern
human population, some go farther back to our common ancestor with
the Neanderthals, and some stretch back to our common ancestor with
the apes and beyond. There is, too, another potential complication in
assessing the evolution of mtDNA. Although many of the distribution

patterns we see today appear to be the result of chance, or of historic events such as migrations of females, it is apparent that mtDNA, which does contain some functional genes, can also be subject to the effects of selection.

The changes in living humans compared to Eve's reconstructed mtDNA genome sequence average about fifty substitutions, and the different mtDNA types of modern humans have been divided into *haplogroups*, clusters that share changes in their DNA sequences, and that have descended from female ancestors who first expressed those mutations. The most ancient haplogroup, named L, was the ancestral one derived directly from Eve and is found across the majority of Africa's populations today. L can in turn be divided into subgroups L0–L3, in terms of their order of branching. The most ancient of these, L0, is found in southern and eastern Africa, with its oldest branches among the Khoisan hunter-gatherers of southern Africa. L1 is mainly found in central and western Africa, including so-called pygmy populations of the central equatorial forest, while L2 is the most common in Africa, at about 25 percent, mainly in the west and southeast. The youngest of these major haplogroups, L3, is common across sub-Saharan Africa, especially in the widespread Bantu-speaking popu-lations, and is thought to have originated in eastern Africa. This makes good sense in evolutionary terms, because this region and L3 were prob-ably the main source of the populations who moved out of Africa and founded the non-African haplogroups M and N, which are found across the rest of the world.

Of course population movements, particularly in the last millen-nium, have translocated many lineages far from their places of origin, and a large industry has grown up to help people trace their ancient ancestry through mtDNA. This has proved controversial because your mtDNA ancestry is only a small part of your total genetic ancestry, but along with Y-chromosome DNA for males, mtDNA is very easy to sample, sequence, and track. Yet even when it is tracked back successfully, the results are only as good as the comparative data that are used to "relo-cate" people (or at least those small bits of their DNA) to their original homelands, and many parts of the world, including Africa, are still poorly sampled for DNA. We do know that African mtDNA contains the most ancient lineages and the greatest diversity for modern humans, consistent with Africa being both our place of origin and the region with the largest ancient population size, which was thus able to conserve that diversity.

Mitochondrial DNA has been widely used to gauge ancient population sizes, though estimates of these from genetic data are fraught with difficulties, one of which is that calculations generally provide an *effective population size*—in essence, the size of the breeding population. For mtDNA, this is the estimated size of the pool of "mothers," while the actual population size (including breeding males and individuals either too young or too old to be involved in breeding) would obviously be much larger. However, many estimates of ancient population size, whether from mtDNA, Y-DNA, X-DNA, or other autosomal DNA, are startlingly low when we consider the billions of humans on Earth today. The long-term effective size of the ancestral population for modern humans might have been only about 10,000 breeding individuals, while the effective size of the female population, judged from surviving mtDNA, is sometimes estimated at less than 5,000!

If such numbers are a true reflection of the original population size in Africa, humans were present only in numbers comparable to those of gorillas and chimpanzees, species that inhabit relatively small parts of the African continent today. Our ancestors cannot have been widespread across the continent, let alone spread far outside of it, but were probably concentrated in pockets, and those pockets would have been vulnerable to extinction. Using three complete human genomes, the geneticists Chad Huff, Lynn Jorde, and their colleagues made comparisons that reached even deeper back in time to suggest that human population numbers a million years ago (the time of *Homo erectus*) were somewhat larger, closer to 20,000 breeding individuals, but even this size could hardly have spread across a continent as large as Africa.

Mitochondrial DNA can also be used to track population growth, and some studies suggest that while haplogroups L0 and L1 grew steadily in their early history, L2 expanded only quite recently, while L3 grew rapidly about 70,000 years ago. In mtDNA terms, as we have seen, the latter group was ancestral to lineages M and N found outside of Africa, so that expansion might well have spilled over into western Asia and hence to the rest of the world.

MtDNA has been used to calibrate events in human evolution, as we saw from the original calculation of Eve's antiquity of about 200,000 years, and from the estimate of the expansion of haplogroup L3 at about 70,000 years, but as with population size estimates such calculations are reliant on several assumptions and can only be approximate. For example, most calibrations are based on the assumption that we split from our closest living relatives, chimpanzees, about 6 million years ago. The number of substitutions in our mtDNA compared with that of chimpanzees is then compared with the number of substitutions determined for other events, such as our split from Neanderthals or our exit from Africa. The ratio of substitutions found is then converted into a "date," along a 6-million-year time scale. However, when substitution rates are determined in very recently diverged human mtDNA, such as in historic populations on islands, or family studies where there are unusual mitochondrial diseases, the rates are much faster than the rate found when comparing our mtDNA with that of chimps. Scientists have argued that "purifying selection" removes many disadvantageous mtDNA mutations through time, thus explaining the rate discrepancy between the short-term and long-term evolutionary events. But when we attempt to calibrate relatively recent events in human evolution, such as the date for Eve or our exit from Africa, should the slow (long-term) rate be used, as it most often is, or should a faster rate be applied?

I recently collaborated with the geneticists Phillip Endicott, Simon Ho, and Mait Metspalu to compare two existing calibrations for recent human evolution with newly calculated substitution rates that are not based on the ancient and somewhat uncertain 6-million-year separation time for chimps and humans. The new rates gave younger estimated ages for recent events in human evolution, but ones consistent with the latest fossil and archaeological data for the exit from Africa and for our arrival in Asia, Australia, Europe, and the Americas. "African Eve" would have lived about 135,000 rather than 200,000 years ago, the exit from Africa would have taken place about 55,000 years ago, and the arrival in the Americas at about 14,000 years ago. If they are correct, these new and younger dates for human mtDNA evolution necessitate rethinking the mtDNA time scale for several key events in our evolutionary history, implying a younger date for our divergence from Neanderthals, a separation of many millennia between the first modern human fossils in

Africa and Eve, and they also cast doubt on ideas of an early exit from Africa toward China and Australia. I will return to these issues shortly, but it certainly seems that geneticists need to reconsider their reliance on the human–chimp divergence to calibrate much more recent events in human evolution.

Compared with mtDNA, the Y-chromosome—the source of data on male history—has been slower to make an impact on the reconstruction of modern human origins than its female-tracking mitochondrial equivalent. One of the major reasons for this is that the Y is actually rather small and boring in terms of its genes and DNA compared with other more enlightening parts of our genome. It is predominantly made up of less informative junk DNA, and only small parts of its genetic material are ever exchanged with the X-chromosome. Nevertheless, it has now been completely sequenced, and increasing refinements in analysis have meant that even this recalcitrant chromosome has yielded important data on recent human history.

The most recent detailed comparisons of the human and chimp Y-chromosomes by Jennifer Hughes and David Page showed that these two are surprisingly different, with the human Y retaining many more coding regions. Because it is inherited through males only, there is a theoretical "Adam" to represent the last common ancestor of all modern Y-chromosomes, and as with mtDNA there is so far no evidence of a more ancient surviving variant of Y that could have been inherited from archaic people like the Neanderthals. Until recently "Adam" was estimated to have lived about 80,000 years ago, much later than "Eve," with the initial and deepest two branches of the Y evolutionary tree widespread in Africa, one common from Bushman to Sudanese populations, the other in central African "pygmy" tribes. But new analyses by the geneticist Fulvio Cruciani and colleagues have instead placed the common ancestor at about 142,000 years ago, most likely (based on present distributions) in central or northwestern Africa. As with mtDNA, populations outside of Africa have lower diversity, this time with a slightly younger common male ancestor some 40,000 years old. Y is also useful in tracking unusual demographic events involving males in recent human history, such as the dominance of one Y-chromosome type across much of central Asia with an antiquity of about 1,000 years—perhaps the legacy of Genghis Khan's habit of impregnating large numbers of women in conquered populations, as well as the historically documented reproductive success of his known male descendants.

The use of autosomal DNA to study human population relationships has a long history, at least in terms of the study of the geographic distribution of its products such as blood groups, proteins, and enzymes. In the 1970s attempts were made to reconstruct the genetic history of humans by combining data on the frequency of many different genetic markers in populations from across the world. However, these often gave conflicting signals, sometimes relating the populations of Europe and Asia together, and sometimes indicating a closer relationship between Africa and Europe. One exception was the pioneering use of a genetic distance technique by the geneticists Masatoshi Nei and Arun Roychoudhury that allowed them to calculate that modern humans were closely related to each other but that Europeans and Asians had diverged about 55,000 years ago, while their ancestors had diverged from Africans about 115,000 years ago.

These estimates look crude now, and no one would suggest that these were real evolutionary "splits," but the inferred relationships were in line with those determined by mtDNA and several other analyses a decade later. The arrival of techniques that used enzymes to chop the DNA into studiable segments (*Restriction Fragment Length Polymorphisms*) led to examination of the gene for betaglobin (which makes up part of our blood's hemoglobin) in 1986, and to early support for the concept of an African origin and a subsequent Out of Africa dispersal. Since then, hundreds of studies of autosomal DNA have shown the same pattern: African populations have the greatest diversity, and people outside of Africa are essentially a subset of that variation. In one of the largest recent investigations of over 1,000 genetic markers in 113 African populations, it was shown that they could be classified into fourteen groups, closely matching known cultural and language affiliations. Populations such as central African "pygmies," hunter-gatherers such as the Sandawe and Hadza of Tanzania, and the Khoisan of southern Africa shared ancestors about 40,000 years ago. What was also interesting was that the latter three populations all speak "click" languages, suggesting that this could have been an ancient shared aspect of their languages.

While autosomal DNA studies have repeatedly confirmed the low diversity of most gene systems in non-Africans, they have also thrown an intriguing light on the pattern of dispersal of modern humans from their ancestral homeland. Just as non-African DNA variation can be seen as a subset of African variation and was originally sampled from it, so, as modern humans dispersed, that pattern seems to have repeated itself over and over again. The front line of expanding moderns from Africa was evidently small in number, and thus these pioneer groups radiating out from southwest Asia themselves only represented a small part of their parent population, with consequent lower DNA diversity. As a relic of that process today, DNA diversity steadily declines with overland distance from Africa, reaching its lowest points in faraway regions such as Arctic Europe, the Americas, Polynesia, and Australasia—and a matching pattern can even be found in the DNA history of *Helicobacter pylori*, a bacterium that infects most of us and can cause peptic ulcers!

Just as intriguingly, this pattern of decreasing diversity from Africa can be picked up in the measurements of skulls of populations from different parts of the world, suggesting that most of the regional differences between crania that are utilized by forensic programs were generated by drift rather than natural selection. I say *most* because there is evidence that certain populations like the Siberian Buryats and Greenland Eskimo underwent head and face shape selection under the impact of extreme cold—being large-headed and flat-faced seems to be advantageous under such conditions. But they represent exceptions to the general rule. Such decreasing diversity in both genes and morphology provides a challenge to the assimilationist idea that the expanding moderns mixed everywhere with remaining populations of archaics such as the Neanderthals and descendants of *Homo erectus* in the Far East. If that were so, we would expect to see repeated reversals of the decline in diversity where such different kinds of humans had significant input into modern human variation, and this has not been observed so far—with one very important exception. The geneticist Jeffrey Long and his colleagues recently reported a hot spot of increased diversity in the southeast Asian islands of Melanesia, and this was a clue to significant local complications in the Out of Africa dispersal, as we will see shortly.

One of the most difficult aspects for some people to accept, if we

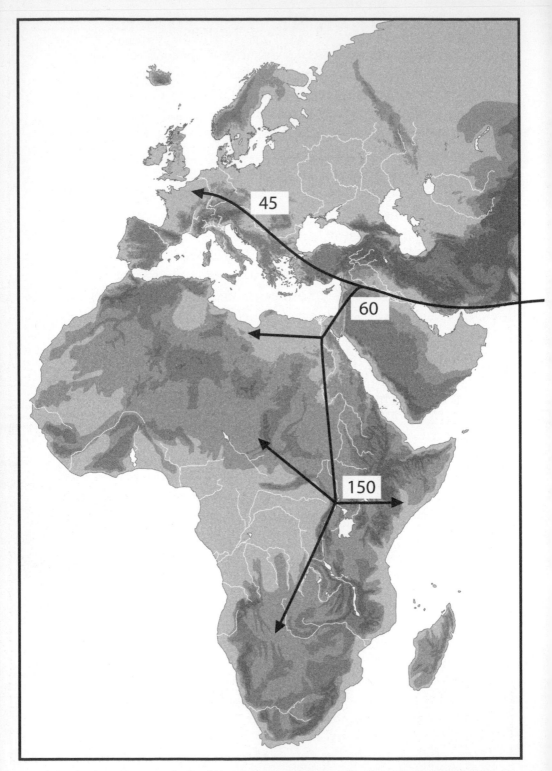

Map showing the spread of early modern groups as traced using mitochondrial DNA (numbers refer to thousands-of-years-ago). The routes are notional, not precise.

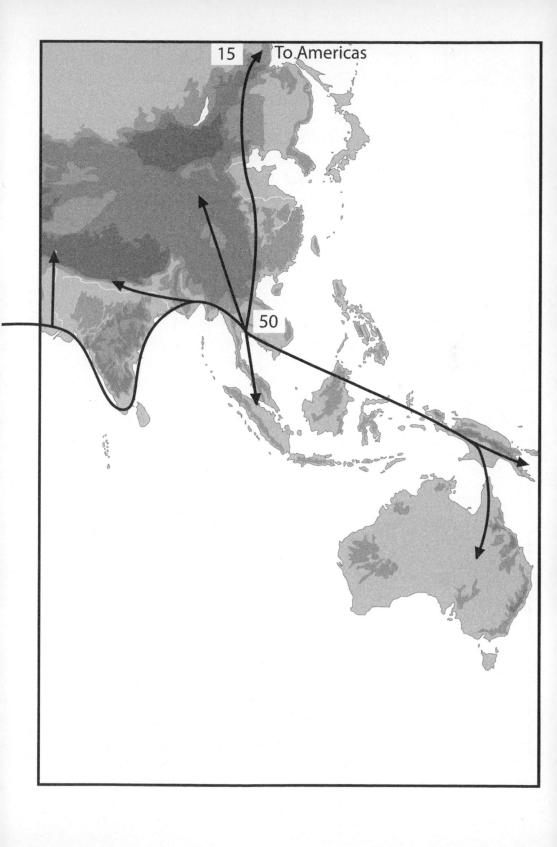

evolved very recently in Africa, is why we all look so different. As I said more than twenty years ago, "we are all Africans under the skin," and yet what lies in and on the skin seems to distinguish us from each other so markedly. Humans come in many different sizes, shapes, and colors and differ in the form of their eyes, hair, nose, and lips. These "racial" or, better, regional or geographic differences are immediately apparent, and thus some people assume they must be highly significant genetically. Yet if we had a recent African origin, these differences must have evolved after we became the modern human species and started to spread out from our place of origin. Thus we evolved our shared species-specific features—our high and rounded skull, small brow ridges, small retracted faces, chins, and so on—first in Africa. Then, on that shared modern template, the regional features were superimposed. But what led to those additions? Here there are several different ideas, and two in particular stand out: climatic adaptation through natural selection, and sexual (in humans, also cultural) selection. Surprisingly, despite Darwin's (and Wallace's) emphasis on natural selection as the predominant agent of evolutionary change, when Darwin came to publish *The Descent of Man* in 1871, it was the second part of the title—*and Selection in Relation to Sex*—that dominated his thoughts on the evolution of "racial" characters in humans.

> We have now seen that the external characteristic differences between the races of man cannot be accounted for in a satisfactory manner by the direct action of the conditions of life, nor by the effects of the continued use of parts, nor through the principle of correlation. We are therefore led to inquire whether slight individual differences, to which man is eminently liable, may not have been preserved and augmented during a long series of generations through natural selection. But here we are at once met by the objection that beneficial variations alone can be thus preserved; and as far as we are enabled to judge . . . none of the differences between the races of man are of any direct or special service to him . . .
>
> We have thus far been baffled in all our attempts to account for the differences between the races of man; but there remains one important agency, namely Sexual Selection, which appears to have acted powerfully

on man, as on many other animals . . . it can be shewn that it would be an inexplicable fact if man had not been modified by this agency, which appears to have acted powerfully on innumerable animals. It can further be shewn that the differences between the races of man, as in colour, hairiness, form of features, &c., are of a kind which might have been expected to come under the influence of sexual selection . . .

For my own part I conclude that of all the causes which have led to the differences in external appearance between the races of man, and to a certain extent between man and the lower animals, sexual selection has been the most efficient.

While I think Darwin was right to question natural selection as the factor behind features like thick or thin lips, and the distinctive eye form of many oriental populations, we saw in chapter 4 that Allen's and Bergmann's "rules" of climatic adaptation seem to affect body-shape variation in humans from different regions. It seems likely that nose shape and skin color have been shaped by natural selection—in the former case via differences in the local temperature and humidity of the air, and in the latter case through the strength of sunlight, particularly in ultraviolet (UV) wavelengths. The theory behind skin pigmentation differences is that they have evolved as a balance between the need for the skin to receive enough sunlight to allow essential vitamin D to be synthesized under our skin, and the need to protect our skin from an excess of UV, which can damage folic acid levels (vital during pregnancy) and skin cells, leading to cancers. Studies are complicated because humans have recently become much more mobile, thus confusing some correlations of pigmentation with UV levels that may have existed previously.

Nevertheless, there seems to be clear negative evidence for the protective benefits of dark pigmentation in the prevalence of folic acid destruction and skin cancers in lightly pigmented people of European origin who moved to high UV regions such as South Africa and Australia. And, demonstrating the opposite process in highly pigmented peoples, African and southern Asian peoples moving to northern regions such as Scotland and Canada have a greater risk of vitamin D deficiency (and thus of the disease of rickets), which is exacerbated if they also go out less and cover their bodies more when they do go out.

These data imply that our original (African) ancestral skin color was indeed darkly pigmented, and that selection favored lighter skins as modern humans spread to regions where UV levels were low and the diet was not providing enough vitamin D. In fact the favored mutations that produce lighter skin in Europeans are young (on some estimates one of the most important genetic changes only occurred about 11,000 years ago), and several (but not all) are different from those that have evolved recently in north Asians. But this is not to say that natural selection is the only factor at work in human skin color, or nose shape for that matter, since sexual/cultural selection could also have played a part. An example of this comes from blue eye color, which is common in northern Europe. The mutation responsible for this has probably occurred many times by chance in human evolution but has not generally been favored. However, the European version seems to be young—less than 20,000 years old—so we can imagine it originating in a Cro-Magnon population somewhere in Europe. Lightly pigmented eyes are certainly disadvantageous in conditions of strong sunlight, but in Europe the unusual nature of the light color might have led to it being favored as an attractive and only mildly disadvantageous variant, which then proliferated through sexual/cultural selection. This variant is down to only one tiny segment of DNA, which shows how small genetic changes can produce striking differences in appearance.

"Racial" features have largely evolved more recently, through quite small changes in our DNA, but they have a strong impact on us because they affect what we notice when we meet people for the first time: their color, facial appearance, and hair. Because of their importance in signaling, I have no doubt that such traits could have been selected for sexual/cultural reasons through differing norms of attractiveness or to enhance group identity. But also at work as modern humans dispersed quite rapidly from Africa in relatively small numbers would have been the effects of *drift* and *founder effect*. The former process is the result of random events; once populations stop exchanging genes, they may "drift" apart purely by chance. The latter process is the result of chance too, but in this case a small and perhaps atypical group may go on to found a much larger population, which will reflect their idiosyncratic genetic makeup rather than that of the original. These phenomena may have combined, as mod-

erns spread rapidly, to produce something called *surfing*, after the popular water sport: particular gene combinations that were rare can end up as
very common if they are lucky enough to "ride" on the expanding population wave, and hence proliferate in the new daughter populations—and
this certainly seems to explain some distinctive gene frequencies outside
of Africa.

These genetic complexities show why old "racial" categories such as
"Negroid," "Caucasoid," "Australoid," and "Mongoloid" have largely been
abandoned by science, because they are not meaningful descriptors of levels of biological variation. Additionally, all of us are to a greater or lesser
extent "mixed" in our origins, since each of our genes will have its own
separate history, and they will not all tell the same story of origin. Hence
the golfer Tiger Woods reacted to being hailed as a model for blacks in
America (a rather tarnished one, now) by saying he was actually Cablin-
asian, as in Caucasian-Black-[American] Indian-Asian, reflecting his
multiple lines of descent. As we said, African populations probably contain as much genetic variation as the rest of the world put together, and the
boundaries between these categories are often fuzzy in reality. This is not
to say that many populations cannot be distinguished at a general level by
the prevalence of common inherited features, and this is also reflected in
traits like cranial and facial shape, which is why forensic scientists can
often confidently place a skull back into its parent population through
study and measurement. But in line with the expectations of a recent African origin, if we try those forensic tests that are based on modern patterns
of regional variation on early modern skulls more than 20,000 years old,
the results are invariably confused. Hence when I tested the 30,000-year-old
Předmostí skulls from the Czech Republic, they came out as "African,"
while one of the Upper Cave Skulls from Zhoukoudian, China, appeared
"Australian." This does not imply a close relationship to those modern
populations, but rather that a kind of regionality existed then that was different from the pattern we have today.

The hoary subject of apparent differences in brain quality and IQ
between regional populations is not something that is going to go away
any time soon. In this respect, things have not changed much since I was
threatened with legal action over what was written about the subject in
one of my previous books, *African Exodus*. I don't intend to say much

more about this controversial subject here, except to acknowledge that some cognitive differences could, of course, have evolved over the last 50,000 years (for example, see the discussion of the microcephalin gene later in this chapter), just as they have in physical features. But if so, I would expect a large and genetically varied region like Africa to show a high level of such differences, rather than the supposed uniformly low IQ values that some studies report. Additionally, as other research has shown, IQ tests only measure some aspects of "intelligence," and environmental differences in nurture, nutrition, and health make a strong contribution to the results too.

Having looked at the genetic variations between us and our closest living relatives, the chimpanzees, and then at variation within our own species, we now turn to the tremendous breakthroughs that have been made in studies of the DNA of our close extinct relatives the Neanderthals. Twenty years ago, the idea that useful genetic data could be recovered from Neanderthal fossils to compare with our own sounded like science fiction, given the huge problems of extracting minute traces of DNA from ancient bones that had suffered from the effects of degradation, water, temperature changes, and soil acids for many millennia. Even if it was preserved (which seemed unlikely), it would be too difficult to find, too difficult to recover in large enough quantities to study, and too problematic to distinguish from all the other contaminating DNA that would also be there.

However, the field of ancient DNA did get off the ground in the early 1980s, with the sequencing of part of the mtDNA genome of the quagga, a recently extinct close relative of the zebra, whose skins survived in museum collections. And in 1984 a technique called the *polymerase chain reaction (PCR)* was discovered, which enabled researchers to produce millions of copies of specific DNA sequences in just a few hours. With this and improved recovery techniques and comparative DNA databases, it started to become possible to recognize and distinguish ancient DNA, where it survived in sufficiently large and well-preserved quantities. So in 1997 the recovery of the first Neanderthal mtDNA from the most famous representative of the group—the 1856 Neander Valley skeleton—caused a sensation. I was lucky enough to be asked to talk about the research at the press conference in London where Svante Pääbo announced the results, and I remember getting so carried away that I hailed it as an achievement comparable with landing someone on Mars! But for paleoanthropology it was a remarkable breakthrough, although things have moved on so fast in the last decade that over twenty Neanderthal fossils have now yielded this genetic material.

Because our cells generally contain hundreds or thousands of copies of the mtDNA genome, compared with the single set of autosomal DNA contained in each nucleus, and because the mtDNA genome was completely known by 1981, mtDNA was specifically targeted in early research on ancient DNA. But by 2006, using particularly well-preserved Neanderthal fossils and massive improvements in analytical techniques and computing power to recover and recognize small ancient DNA fragments, two international teams of scientists reconstructed the first large-scale genetic maps of the Neanderthal autosomal genome. Two fossil sites have proved particularly valuable in Neanderthal genome work, and in both cases their human remains may have resulted from cannibalism; in fact, there is speculation that defleshing the bones may even have helped the preservation of ancient DNA by heading off some of the proximate causes of DNA decay. One is the cave site of Vindija in Croatia, where small fragments of leg bones have by far and away the best preservation of Neanderthal DNA found so far, and the other is El Sidrón in Spain, which we discussed in chapter 4, and where great attention has been paid to recovering fossils with the minimum possibility of contamination by recent DNA.

The 454 Life Sciences company recently developed new instruments that allowed around 250,000 DNA strands to be sequenced in about five hours on one machine, and thus running several machines in tandem gave phenomenal improvements in recovery and recognition of the 3 billion pairs of chemical bases that originally made up the genome of these Neanderthal individuals. The 454 technique uses "shotgun" sequencing, in which DNA is chopped into huge numbers of short segments, and it is thus ideally suited for the tiny fragments of nuclear DNA required for ancient genome reconstruction. The old PCR technique was really only suitable for looking at longer fragments, such as in Pääbo's early Neanderthal mtDNA work, but a development by the researcher Paul Brotherton and his colleagues called *SPEX* (*single primer extension*) now also holds great promise for the recovery of small fragments of the Neander-thal genome, in a more targeted approach than that of 454 analysis.

Genomic DNA in one of the El Sidrón individuals and another Nean-derthal from Monti Lessini in Italy is providing some of our first glimpses of the constitution of southern European Neanderthals. They had muta-tions in the structure of a pigmentation gene, MC1R, which would have been expressed in red hair and pale skin, and despite the media reeling out a number of celebrities and sports stars with "ginger" hair, saying they were all "Neanderthals," the more interesting story is that the Neander-thal variant was, in fact, distinct from that found in people of European descent today. Lighter pigmentation in humans has probably evolved for several reasons, but these include facilitating the synthesis of vitamin D in our skin under northern conditions of reduced sunlight. That at least some Neanderthals evolved their own depigmentation is not surprising when we consider that they were living in Europe for hundreds of millen-nia before modern humans did. What is surprising, though, is that if there was significant interbreeding between moderns and Neanderthals in Europe, potentially advantageous genes for lighter skin did not spread from them to us—and other research suggests that some of the gene vari-ants that produce the light skin of many Europeans are probably less than 15,000 years old.

At least one of the El Sidrón Neanderthals had mixed genes at the TAS2R38 site, which in modern humans controls an ability to taste (or not taste) the bitter chemical phenylthiocarbamide (PTC). Related chemicals occur in leafy vegetables like Brussels sprouts and cauliflower, as well as in some poisonous plants, and it is possible that the tasting/nontasting dichotomy had evolved in more ancient humans as part of a balance between nutritional needs and detecting the danger inherent in some bitter and poisonous plants. At least two El Sidrón individuals also shared another genetic system with some of us, in the form of blood group O, which is coded on chromosome 9. The famous ABO blood groups are distinguished by the presence or absence of particular antigens on the sur-face of red blood cells, which give resistance to different diseases. In blood group O, a mutation blocks the action of an enzyme that produces the A and B antigens, and while this might seem to be disadvantageous, some disease agents actually lock on to the antigens, so that lacking them can confer an advantage. Chimps also have the ABO system, although group O is less common in them, so it seems likely that the system is a shared inheritance between chimps, us, and Neanderthals, with different diseases constantly pruning the patterns of individuals with the most vulnerable blood types. As more Neanderthals are sequenced, we will be able to compare their frequencies with those of humans today.

Carles Lalueza-Fox and his colleagues ingeniously also used the circumstances of the El Sidrón site, where a possible Neanderthal family group had become fossilized, to provide a glimpse of their social structure. The three men had identical mtDNA sequences, while the three women each had different sequences (but were related to three of the children), so if this really was a family group, it implied that the males were closely related and had probably stayed in their natal group, while the females had joined from other bands. Such exchanges of mates (perhaps mostly, but not always, peaceable) are important in reducing inbreeding and suggest that females were the predominant agents of gene flow, and perhaps also of any cultural transmission, between Neanderthal groups. This social system is known as *patrilocality* and is the most common in modern hunter-gatherers, and seems to be yet another behavior shared between Neanderthals and moderns.

A more controversial finding in two of the El Sidrón people was the presence of a gene also found in modern humans called FOXP2, which has misleadingly been called "the language gene"—as though only one gene is likely to be involved in this very human faculty. In fact this developmental gene became known through a kind of reverse engineering, because when it malfunctions in humans it leads to inhibitions in the comprehension and production of language, both in brain pathways and in the physical control of muscles concerned with the production of speech. When the gene was sequenced and compared between other primates and humans, it was discovered that there were two unique mutations in the human version, which had presumably been selected to help facilitate our power of speech. Further research showed that it is at work in several areas of the brain concerned with cognition and language, and the human version of FOXP2 regulates (amplifies or moderates) the activities of more than a hundred other genes, whereas the "ancestral" version found in chimps has no such effect.

Our special version of the gene is not just about language, but it certainly does seem to be implicated in establishing neural pathways and the anatomical structures for speech. Thus there was speculation about whether Neanderthals would have possessed these same mutations, or, if lacking them, may also have lacked the capabilities of speech. The first drafts of the Neanderthal genome seemed to show the presence of the human form of FOXP2, but there were concerns even among the research team that this might be as a result of contamination from recent human DNA. However, the discovery of the "advanced" version in the carefully screened El Sidrón individuals seems to confirm its presence in these Spanish Neanderthals. So does this mean the Neanderthals must have had fully modern language? In my opinion, it does not, any more than the fact that the hyoid bones which sat in their throats were similar in shape to our own. But what is indicated is that we have no reason—from these elements of their biology—to deny them the potential for modern human speech capabilities. Whether they actually had our language abilities would also have depended on their own evolutionary pathway in behavioral complexity and the structure of the brain and vocal apparatus, as well as any evolutionary constraints that might have been at work from their distinctive anatomy.

An even more controversial issue than the presence or absence of the "modern" version of the FOXP2 gene in Neanderthals is whether they had a particular version of the MCPH1, or microcephalin, gene. In another case of reverse engineering, the action of this gene in humans became known through occasional failures in fetal development, where mutant versions seemed to be related to microcephaly (having an abnormally small head and brain). In such cases the faulty microcephalin gene apparently interfered with instructions for the production of neurons in the forebrain, leading to later deficits in the cerebral cortex. There are two main variants today, one most common globally (type D) and the other prevalent in sub-Saharan Africans ("non-D"). The genetic history of these two types appears to be quite distinct: while non-D seems to have developed in Africa and spread from there with the dispersal of modern humans, D has only proliferated in modern humans in the last 40,000 years, suggesting it has been selected as advantageous in at least some regions or situations. Yet the mutations in the genes show that these two types of microcephalin have deep and separate common roots going back over a million years, so where could the "new" D variant have come from? Attention focused on the Neanderthals as a possible source, implying that modern humans outside of Africa could have acquired their "young" variant from the Neanderthals, but sadly for this hypothesis, genome sequencing so far has shown that the Vindija Neanderthals possessed the ancestral "African" version of the microcephalin gene. Moreover, further research cast doubt on the whole scenario by failing to confirm the hypothesis that the microcephalin gene is strongly implicated in brain development, quality, and intelligence in "normal" humans. But the case of microcephalin does raise the issue that while mtDNA, Y-chromosome, and most autosomal DNA strongly support a recent African origin and subsequent dispersal for our species, there are variants of some genes that suggest a more complex evolutionary history for *Homo sapiens*.

Just as recent human DNA has been used to estimate past population numbers for *Homo sapiens*, so the small amounts of Neanderthal DNA recovered so far have also been subject to similar analyses, with clear and rather negative implications for Neanderthal viability. The complete reconstructed mtDNA genomes of six Neanderthals from Germany, Spain, Croatia, and Russia differ at only fifty-five locations out of a total of more than 16,000 base pairs, which is far less mtDNA diversity than in modern humans, and only a tiny fraction of the variability found in great ape species today. Estimates of population size from these data put the effective population size of Neanderthals across Europe and western Asia as low as 3,500 breeding females, although, as we have seen, that could translate into a much larger total number of people. In addition, they seemed to harbor a relatively greater number of potentially damaging mutations, which could have affected the structure of their proteins, something that often comes with smaller population sizes. Given that these were late Neanderthals sampled across much of their wide range, we can see how they could have been a threatened species even without the destabilizing impact of the arrival of modern humans in their home territories.

The partial Neanderthal genomes produced in 2006 contained some contradictory data, and doubts were soon expressed about whether the results were affected by remaining contamination from modern human DNA. Further investigations showed that this was indeed so, perhaps as much as 15 percent in certain areas. But now a composite and nearly entire Neanderthal genome has been drafted, providing rich data that promise yet more insights into their biology, from eye color and hair type through to brain quality and language skills. An international team of more than fifty researchers reconstructed more than 3 billion bits of DNA coding, again predominantly from three small fragments of bone from the Croatian cave of Vindija. These represented female Neanderthals who died around 40,000 years ago, and they have now been immortalized through their DNA. The results still largely confirmed the Out of Africa thesis, the overall distinctiveness of the Neanderthals, and a separation time from our lineage of about 350,000 years. But when the new Neanderthal genome was compared with those of modern humans from different continents, the results produced an intriguing twist to our evolutionary story because the genomes of people from Europe, China, and New Guinea lay slightly closer to the Neanderthal sequence than did those of individuals from Africa. Thus if you are European, Asian, or New Guinean, you probably have a bit of Neanderthal in your makeup.

One explanation is that the ancestors of people in Europe, Asia, and New Guinea interbred with Neanderthals (or at least with a population that had a component of Neanderthal genes) in North Africa, Arabia, or the Middle East as they exited Africa about 60,000 years ago. That ancient human exodus may have involved only a few thousand people, so it would have taken the absorption of only a few Neanderthals into a group of *Homo sapiens* for the genetic effect—greatly magnified as modern human numbers exploded—to be felt tens of thousands of years later. The amount of Neanderthal genetic input is estimated to be about 2 percent overall, a surprisingly high figure to me and other adherents of Out of Africa, who thought that any slight traces of interbreeding would have disappeared in the intervening years. Moreover, subsequent examination of six thousand worldwide modern samples by geneticist Vania Yotova and colleagues has revealed that non-African X-chromosomes have as much as 9 percent of Neanderthal-derived DNA in one particular location. What any of the shared DNA does for us, if anything, remains to be determined, but that will certainly be a focus for the next stage of this fascinating research. Alongside the apparent transfer of Neanderthal DNA into some of us, the comparisons also revealed more than two hundred genetic changes that we share to the exclusion of Neanderthals and chimpanzees. Some of these are in genes involved in brain functions, the structure of the skull and skeleton, the skin and its associated organs (such as hair and sweat glands), energy functions, and sperm activity.

These breakthroughs come at a time when renewed claims have been made that Neanderthals and early modern humans (Cro-Magnons) interbred in Europe about 35,000 years ago. Both fossil and DNA data indicate that the Neanderthals were a distinct lineage from modern humans, but a closely related one, and, as I explained, the level of morphological difference in the skeleton is comparable to that in recent primates and fossil mammals that demarcate distinct species. However, closely related mammal species may still be able to hybridize, so this was certainly possible between Neanderthals and Cro-Magnons.

The anthropologist Clifford Jolly, who was my first teacher in paleo-anthropology at University College, London, has made a special study of baboons and their relatives in Africa today, and these monkeys seem to represent distinct species groups in their appearance and behavior, yet when their DNA is analyzed, it is apparent that these "species" often exchange genes on at least a small scale, where they overlap geographically. As he said with reference to fossil human species such as the Neanderthals: "The message is to concentrate on biology, avoid semantic traps, and realize that any species-level taxonomy based on fossil material is going to be only an approximate reflection of real-world complexities." I think we should certainly remember those wise words before we make any absolute statements about what might or might not have taken place if and when our forebears met the Neanderthals.

The essential question with regard to the behavior of our ancestors and the Neanderthals is, of course, did they regard each other as just another group of people? We don't know the whole story, and the answer may have varied from one time and place to another, especially given the vagaries of human behavior. I take a different view on this than my friend Erik Trinkaus, who sees Neanderthal input in most of the earliest moderns in Europe, for example, in a child's skeleton from the Lapedo Valley in Portugal. This fossil was buried with the red ocher and grave goods typical of many Gravettian burials (as discussed in chapter 4) about 27,000 years ago, and it has been described in detail, with contradictory indications. Nearly everything in its anatomy suggested that this was a fairly typical Cro-Magnon child, but the robusticity and proportions of its limbs and some of its dental features suggested to some that it represents evidence for admixture between Neanderthals and modern humans. Given its Gravettian age, apparently several millennia after the Neanderthals had vanished, the inference seems to be that this is a Cro-Magnon child that has retained some Neanderthal genes and characteristics from an earlier phase of interbreeding. However, in this and other cases, rather than perceiving features that definitely came from Neanderthals, I see some that were present in the ancestors of these modern people in their place(s) of origin, or that represent individual variations that overlap with those of the Neanderthals in some respects. When we have a reasonable sample of North African or Asian early moderns dating from about 50,000 years—the same period as many of our Neanderthals—we will be able to see what their morphology was, and we will be better able to determine whether features could have come from Neanderthal admixture or were due to ancestry in a different region.

Many years ago I remarked that the one thing you might have expected the Neanderthals to bequeath to the Cro-Magnons in any interbreeding was physical (in terms of body shape) and physiological adaptation to the cold, whereas the reality is that the early Cro-Magnons had completely contrasting linear physiques, while Europeans of today have poor cold tolerance compared to many other modern human groups—hardly what you would expect if their ancestors included Neanderthals. This evidence against interbreeding is seemingly duplicated when it comes to skin color, as we already saw, since modern Europeans evolved their own lighter pigmentation rather than borrowing that of the Neanderthals. But if genome comparisons show that there was interbreeding, why didn't these apparently useful features transfer across?

I think the answer may come from recognizing that the place everyone has focused on when thinking about interbreeding is Europe, at the time of the last Neanderthals. But by then they were a dying breed, few in number and with low diversity. If the interbreeding actually happened earlier, in a warmer region or a warmer period, maybe the Neanderthals involved were not light-skinned and cold-adapted European examples. In fact, the interbreeding might even have happened when people like those from Skhul-Qafzeh and Tabun were in the Middle East 120,000 years ago. If a thousand of those early moderns mixed with just fifty Neanderthals and then survived somewhere in Arabia or North Africa, could they have subsequently interbred with the Out of Africa emigrants 60,000 years later and passed on their hidden component of Neanderthal genes?

However, there is yet another possibility. Ongoing research suggests that the 2 percent or so of Neanderthal DNA in French, Chinese, and New Guinean samples is not the *same* 2 percent, and modeling Neanderthal-modern interactions has led geneticists Mathias Currat and Laurent Excoffier to propose yet another scenario—that this 2 percent or so in the different populations is instead a measure of how ineffective hybridization was in *separate* interbreeding events. It thus indicates a natural limit on the process, caused by biological, social, or demographic factors.

Given our possible role in their demise, should we reverse the process of extinction and attempt to clone a Neanderthal from its newly reconstructed genome? This is something I would have dismissed as pure science fiction only a few years ago, but with the staggering progress recently in genomics no one should rule out an attempt in the future. What I am sure about, though, is that it would be quite wrong to resurrect long-extinct species purely to satisfy our curiosity about them, especially if they were human. Neanderthals were the products of a unique evolutionary history in Eurasia that lasted for several hundred thousand years, but they are gone, along with the world in which they evolved, and we should let them rest in peace.

The breakthroughs in reconstructing Neanderthal genomes were mirrored across Asia in equally remarkable work on what has become known as "Lineage X," or the "Denisovans," by David Reich and his colleagues. A fossil finger bone about 40,000 years old from Denisova Cave in Siberia, which could not be assigned to a particular human species, first yielded a surprising mtDNA sequence, neither Neanderthal nor modern. In fact its mtDNA was more different than that of Neanderthals and moderns, suggesting an origin time well over 500,000 years ago. This finding was backed up when a massive molar tooth from the same levels was shown to possess a very similar mtDNA pattern. But even more remarkably, autosomal DNA was then recovered from the finger bone, allowing parts of the whole Denisova genome to be reconstructed.

Comparisons with the genomes of chimps, Neanderthals, and various modern humans produced even more surprises: Denisova was probably a derivative of the *heidelbergensis* lineage, but one that remained genetically closer to Neanderthals than to modern humans, perhaps indicating intermittent gene exchanges across central Asia where Neanderthals and Denisovans encountered each other. Enigmatic Asian fossils dated between 100,000 and 650,000 years ago, such as those found in Narmada in India and Yunxian, Dali, Jinniushan, and Maba in China, have been considered possible Asian derivatives of *H. heidelbergensis*, or relatives of Neanderthals, so they may be candidates for this ancient eastern lineage. In addition, there are fragmentary remains from Xujiayao in north China and Zhiren Cave in south China that are claimed to show both archaic and modern human features, dating from about 100,000 years ago. These fossils are too incomplete to determine their evolutionary status, but they also hint at additional complexity in the story of modern human evolution in China.

But something else even more remarkable has still to be explained properly: the Denisovans are also related to one group of living humans—Melanesians—which may explain Jeffrey Long's hunch, discussed earlier in this chapter, that they contain distinct archaic genes from the rest of us. The most plausible explanation for this is that Denisovans were present in southeast Asia as well as Siberia, and pre-Melanesian populations migrating through the region from Africa interbred with some of these Denisovans, picking up some 5 percent of their genes. That component might represent only twenty-five Denisovans mixing with five hundred pre-Melanesians, but it was sufficient to give the present-day inhabitants of places like New Guinea and Bougainville as much as 8 percent "archaic" genes—a small Neanderthal component they acquired first, probably in western Asia, and a larger Denisovan component they acquired later, on their way to Melanesia. As with the presence of small amounts of Neanderthal genes, there will now be considerable attention to what, if anything, those Denisovan genes might be doing in modern-day Melanesians. For example, could they have picked up useful defenses against some of the diseases endemic to southeast Asia? This is suggested from studies of the immune system by geneticist Abi-Rached and colleagues, who argue that some variants in the HLA (human leucocyte antigen) system of modern populations in Eurasia could have been picked up through inter-breeding with Neanderthals and Denisovans. Attention will now turn to Australia, as this continent has not yet had full genome comparisons with Neanderthals and Denisovans, but its early colonization and subsequent isolation will make it an important test bed for further hybridization studies.

With such successes in recovering ancient DNA from Neanderthals and Denisovans, unimaginable even fifteen years ago, it may seem strange that there have not been a plethora of papers comparing ancient DNA from archaic populations in many other regions. But the reality is that tropical and subtropical conditions with high temperatures or humidity, or both, severely affect ancient DNA preservation, which is most unfortunate in a case like *Homo floresiensis,* where authentic DNA could have rapidly resolved the fierce arguments about its status as an archaic species or a strange variant of modern humans. While there are hopes for other sites in northern Asia beyond Denisova, and perhaps also for high-altitude locations farther south, it is likely that many ancient human populations will never yield up their DNA for study, and we will be dependent on lateral thinking, such as the study of human parasites, whose DNA provides parallel but independent pathways to the study of our own DNA, and of fossil DNA history (discussed later in this chapter). Non-DNA biochemicals such as bone proteins may also offer better prospects of survival in hostile environments and may yet provide us with useful alternative windows into our evolutionary past. But what of early moderns such as the Cro-Magnons in Europe, who lived under similar conditions to the Neanderthals but even more recently—surely they would be prime candidates for ancient DNA studies? Many of them must certainly contain authentic ancient DNA sequences at least as well pre-served as those of the Neanderthals, and indeed several such sequences have been published. However, in some cases there are lingering doubts about the possibility, even probability, of contamination by recent human DNA, which might be indistinguishable from the authentic material.

Indeed, in 2001 there were claims for the recovery of ancient DNA from Australian fossils dated to between 10,000 and 40,000 years old. Ten out of twelve specimens tested from the Willandra Lakes and from Kow Swamp had apparently yielded human mtDNA sequences, and one of these, from the 40,000-year-old Mungo 3 burial, was claimed to form an outgroup compared with all other fossil (Neanderthal) and recent human sequences. It was claimed that the distinctiveness of the Mungo 3 sequence undermined genetic support for a recent African origin, and instead was confirmation of a Multiregional or Assimilation model of modern human origins. (One of the authors of the study was the multiregionalist Alan Thorne.)

However, critics like myself, Alan Cooper, and Matthew Collins soon pointed out that the claimed recovery rate for the Australian ancient DNA was quite exceptional compared with results from elsewhere, particularly considering that some of the specimens had been buried for millennia under desert sands in conditions of extreme summer heat. Moreover, in some cases the DNA had been recovered from scraps of bone left in storage boxes after the skeletons had, most unfortunately, been reburied or cremated at the behest of aboriginal communities. Furthermore, standard experimental protocols for ancient DNA work had not been employed, suggesting the possibility of contamination, and reanalysis of the published DNA data, using a larger number of recent Australian and African sequences for comparison, demonstrated that the Mungo 3 sequence did not, in fact, form an outgroup to recent human mtDNA; nor did it present any serious challenge to a recent African origin.

Now, reanalysis of the samples using the latest techniques is being carried out, to establish the extent of recent contamination. The lessons learned will be applied to further studies, which are concentrating on distinguishing ancient from recent DNA segments by the signals of damage that they carry. Such work is likely, at last, to provide us with unequivocal samples of ancient DNA from Cro-Magnon and other early modern fossils.

A somewhat surprising—even disturbing—insight into human evolution can also be obtained from the study of some of our fellow travelers: hair and body lice. Lice parasites, which live off human blood, got a mention earlier in research that tried to estimate when humans may have first regularly used the clothing and bedding in which some of them lived. We are the victims of two distinct forms of lice, *Pthirus*, the pubic louse, and *Pediculus*, the head and body louse, and there must be a complex history behind our hosting of these two distinct genera. Our head louse is most closely related to that of chimps, which fits with an evolutionary divergence of our species and theirs about 6 million years ago. Yet strangely, our pubic lice are most closely related to those of the gorilla, but with an apparent divergence of only about 3 million years. This suggests that our head louse diverged with us, as the human and chimp lines underwent their evolutionary separation, but the younger jump of our pubic louse from the less close gorilla lineage must have a different explanation: our ancient African ancestors had direct contact with ancestral gorillas— perhaps sexual, perhaps sociable, perhaps involving conflict or predation. The separate existence of pubic hair in our ancient African ancestors, which presented an opportunity for the transfer from gorillas about 3 million years ago, also implies that we had lost much of our intervening body hair by that time.

The mtDNA of our head and clothing louse comes in three quite distinct lineages, unlike our own mtDNA. The most common group is worldwide in distribution today and shows evidence of an expansion about 100,000 years ago, which fits well with the expansion of modern humans within and then outside of Africa. The second lineage, most common in Europe, diverged from the first about a million years ago, and a rare third lineage found in just a few individuals in Africa and Asia had an even older divergence at around 2 million years. The geneticist David Reed explained that one way to account for the results is to argue that the root population of modern humans was large enough to host these distinct lineages for about 2 million years, but his calculations showed that this was very unlikely.

Another possibility is that there was interaction between human populations that had been distinct for much longer than 200,000 years, which implies contact between modern and archaic humans such as Neanderthals or, as we have seen, the Denisovans. The kinds of interaction cannot be determined, but it could range from interbreeding to contact with bedding, through to aggressive confrontations or even cannibalism, where lice could have jumped from victims to the perpetrators. As an example of the latter, historical studies showed that the Torres Straits islanders, living between New Guinea and Australia, used to keep the heads of both their deceased relatives and their enemies. In the latter case, this sometimes involved eating parts of the face and eyes of the trophy head; such behavior in the past could have allowed the spread of parasites between distinct human populations and even species. If they were able to jump species before their host populations became extinct, these louse lineages literally gave themselves a new lease on life. And of course the transfer of pathogens could have gone in the reverse direction too, and it remains possible that infections of one kind or another brought out of Africa by modern humans contributed to the demise of archaic humans elsewhere.

In this chapter we have discussed the data within our genomes and those of the Neanderthals and Denisovans: genomes that document the evolution of, and at least occasional contact between, these closely related lineages of humans. The evidence confirms that we have a predominantly recent African origin, but worldwide our species is not purely and entirely Out of Africa. Within that continent, our ultimate ancestors were few in number and probably lived in small pockets. Our earlier discussion of the development of behavioral modernity showed its patchy genesis across Africa—a genesis that I compared to brief episodes, like a candle flickering and then being extinguished. So what finally changed to keep that flame burning and then intensified, in order for our species to begin its seemingly inexorable rise to world domination? There are many ideas and theories, and I will start to explore them in the next chapter.

8

Making a Modern Human

AS I EXPLAINED IN THE FIRST CHAPTER, WHEN I BEGAN MY DOC-
toral research in 1970, the origin of modern humans was hardly recognized as a specific topic worthy of scientific study. The standard classification
of humans had living people, the Neanderthals, and diverse remains from
sites like Broken Hill in Africa and Solo in Java all classified as members
of our species. With such different-looking fossil members within *Homo
sapiens*, the origins of features like a chin, a small brow ridge, and a globular skull were, not surprisingly, lost in a morass of diversity. Moreover,
with the predominance of Multiregional or Neanderthal Phase models,
the origins of those features were apparently scattered among many different ancestors living right across the Old World, so modern human
evolution was not so much an event as a tendency; we were merely the end
result of continuing long-term trends in human evolution in features like
increasing brain size and decreasing tooth and face size. For human
behavior too, there was an emphasis on gradual evolutionary trends; for
example, in France the "transition" from the Middle Paleolithic Mousterian to the Upper Paleolithic Gravettian via the Châtelperronian industry
was seen as supporting a parallel local evolution from Neanderthals to
Cro-Magnons.

It looks very different forty years on. For most scientists, Africa has
been established as the center for both our physical and our cultural origins. The evolution of "modern" *Homo sapiens* can be viewed as an

important physical and biological event, backed up by both fossil and genetic evidence. Many researchers would also trace back to Africa the origins of the complex behavior apparent in the Upper Paleolithic figurines and painted caves of Europe. And yet, much as I am delighted with the way the subject of modern human origins has taken off to become one of the most dynamic areas of research in paleoanthropology, I am still puzzled by many aspects of the African origin of our species. When I look critically at what we do know and, more important, what we still don't, I feel we are not yet close to a full understanding of those origins, as I hope to explore in these final chapters.

In the 1980s the issue for people like me, Günter Bräuer, and Desmond Clark was to get people to take the idea of a recent African origin for modern humanity seriously at all, let alone discuss how that origin might have come about. In what was a real struggle against some very influential and at times vitriolic opposition, I'm sure at times we oversimplified both our views and those of the multiregionists, and played down complexity in the data, in what became an increasingly polarized and sometimes bitter debate. At times also, as my views on a recent African origin developed, I favored the idea that our species evolved very rapidly in one small area—a sort of African "Garden of Eden." But the general view has been that there was probably a relatively gradual evolutionary sequence in Africa from archaic humans (*Homo heidelbergensis*, sometimes also called *Homo rhodesiensis*) to our species, *H. sapiens*. *Heidelbergensis* fossils in both Africa and Europe are dated to about 500,000 years old, while, as we have seen, fossils representing our species have been found in Ethiopia at Omo Kibish and Herto, dating to between 160,000 and 195,000 years, with more fragmentary remains from Guomde in Kenya perhaps 250,000 years old. The assumption has been that a gradual accumulation of modern characteristics in Africa would have paralleled a comparable buildup of Neanderthal traits in Europe, from a similar *heidelbergensis* ancestor.

What triggered the evolution of modern humans in Africa, and why this happened at all, is still uncertain. Did social or technological advances promote evolutionary change, or was geographic isolation following severe climatic change responsible? It is not yet even clear where the first "modern" population or populations lived, but the areas of eastern and southern Africa have vied for the title "Cradle of Modern Humanity." The fossil record has highlighted Ethiopia and Kenya in East Africa as the most probable location for our origins, but this is also the region with the best fossil record for the period. In contrast, South Africa has a poorer fossil record but a much richer behavioral one for the Middle Stone Age, which is why some workers claim that region as the real focal point for modern human origins. Recent discoveries also shifted the focus farther north to Morocco, where reevaluation of previous finds and the discovery of new ones suggest that even northwest Africa cannot be excluded as a center for modern human origins. We must also remember that at least 50 percent of African regions that have stone tools from this period have not yielded a single fossil human relic to show us who was making the tools in question. So bearing these points in mind, I would like to take a fresh look at several aspects of our evolution that we already discussed, ranging from biology to behavior, to the role of climate in our evolution, in the hope of throwing further light on our mysterious African origins.

First we'll look at the brain, because one leading theory about our African origins—that of the archaeologist Richard Klein—argues that the development of modern human behavior came about suddenly around 50,000 years ago, as a result of genetic mutations that enhanced the workings of our brain, essentially making us "modern" at a stroke. A similar view is espoused by the neuroscientist Fred Previc, who highlights the importance of the neurotransmitter dopamine to human creative thought and hypothesizes that it reached critical levels about 80,000 years ago, driving behavioral evolution to modernity. Unfortunately it is very difficult to test such ideas properly from the surviving evidence, since while we can make a real or a virtual model of the inside of a fossil skull, that will only reflect the external shape and proportions of the ancient brain that was once inside that skull. Such a model can tell us nothing about the internal workings and wirings of the once-living brain, which would have contained billions of interconnected nerve cells. However, from such data we do know that during human evolution our brains have certainly increased in overall volume relative to body mass (this ratio is known as the *encephalization quotient*, or *EQ*). Early humans had EQs of only about 3.4 to 3.8, and this even included *H. heidelbergensis* individuals, who had human-sized brains but much larger bodies than the average today. More evolved humans such as our African ancestors prior to 200,000 years ago and Neanderthals had EQs between about 4.3 and 4.8, and when we arrive at early moderns such as those from Skhul and Qafzeh and the Cro-Magnons, EQ reaches its highest values at around 5.3 to 5.4.

Since then, *H. sapiens* seems to have leveled off at those values or has even suffered a slight decline in EQ. But in brains, as in many other things, size isn't everything, and we can infer that there must also have been significant reorganizations in the human brain for activities like tool-making and speech. In order to maximize the surface area of the outer-most cortical layer of our brain (the "gray matter," which includes nerve cells and their interconnections), it is complexly folded into convolutions, thus allowing our cortical surface area to be about four times that of a chimpanzee's, matching the increase in overall brain volume. While there have been many careful studies of the impressions of these convolutions of *sulci* (furrows) and *gyri* (ridges) on the inner surface of fossil braincases, such markings are often faint and difficult to interpret. Work done in the last century on the fake Piltdown skull's convolutions found many supposedly apelike features, and yet we now know that the skull in question was actually that of a recent human, so much of that work consisted of wishful thinking or even fantasy—I even compared some of the old work to the pseudoscience of phrenology. Yet another approach to the analysis of ancient brains has focused on changes in the relative proportions of the various components rather than their convolutions, as these can be determined quite well from the preserved inner surface of the braincase or from CT data.

The cortex or cerebrum is by far the largest part of the brain in humans and is divided centrally into two cerebral hemispheres—the left and right—which have different specializations, but which are interconnected by bundles of nerve fibers. The cerebral hemispheres are made up of four lobes, corresponding in position to the cranial bones of the same name: the frontal, parietal, temporal, and occipital lobes. We know quite a lot about the general roles that these lobes play in our brain: the frontal is involved with thinking and planning; the parietal in movement and the senses; the temporal with memory, hearing, and speech; and the occipital with vision. Tucked underneath and behind the cerebral hemispheres is the smaller cerebellum, which is predominantly concerned with regulation and control of the body.

However, recent studies showed that the cerebellum also plays a role in many so-called higher functions and is extensively interconnected with the cerebrum. As well as regulating body functions, it seems the cerebellum is also concerned with the processes of learning. The increase in both gross brain size and EQ really took off about 2 million years ago, soon after the first clear archaeological evidence of both meat eating and toolmaking appeared in Africa. All areas of the brain enlarged, but proportionately the cerebral hemispheres increased more than the cerebellum. The pace of cerebral enlargement accelerated in *heidelbergensis* and peaked in Neanderthals and early moderns, seemingly correlated with an increase in behavioral complexity. But interestingly, in recent humans this long-term pattern has reversed, since the cerebellum today is proportionately larger. At the moment it is not clear what, if anything, this change signifies. On average, human brains have shrunk some 10 percent in size over the last 20,000 years, so is the cerebellum having to maintain its size more than the cerebrum, or, as some have claimed, does a relatively larger cerebellum perhaps provide greater computational efficiency? We simply do not know the answer yet.

It is certain, however, that the overall shape of the brain and the braincase that envelops it changed from archaic to modern humans, becoming shorter and higher, narrower lower down and broader higher up, with a particular expansion in the upper parietal area. Brain shape is inevitably closely matched to skull shape, since the two must develop and grow in harmony—but which is the driver and primary determinant of their closely matched shapes? This is not a simple question to address, since even the braincase and the brain do not grow and exist in isolation. For example, the base of the skull anchors the upper parts of the vocal, digestive, and respiratory tracts and articulates the head and the spine, while the front of the skull contains the teeth and jaws and the muscles that work them. Those factors seem to have constrained brain shape from changing much in those regions.

But the upper areas of the skull and brain are not so constrained, and my Ph.D. results back in 1974 highlighted changes in the frontal, parietal, and occipital bones of modern humans. Each contributed to the increased globularity of the braincase in modern *sapiens*, and our domed forehead is particularly noticeable. There are data from my collaborative research with Tim Weaver and Charles Roseman that suggest that many of these cranial changes might not be significant in evolutionary terms and could simply be the result of chance changes (genetic drift) as modern humans followed their own separate path in isolation in Africa. I will return to this issue in chapter 9, but nevertheless the cranial shape of modern humans is so idiosyncratic, compared with the patterns found in all the other known human species, that I do think it is worth considering whether brain evolution might lie behind our globular vault shape. This is particularly so as research by Philipp Gunz and his colleagues suggests that this shape change began to separate archaic and modern skulls soon after they were born.

It seems likely that important changes occurred in the frontal lobes of our brain, given their importance in forward thinking, yet I was surprised by CT studies of fossil skulls in which I was involved. These showed that the profile and relative size of the frontal lobes inside the brain cavity had changed much less in modern humans, compared with the obvious external changes in forehead development. At the rear of the skull, the occipital bone is relatively narrower and more evenly curved in modern humans. In *erectus* and *heidelbergensis* skulls the occipital was more sharply angled, and this must have been partly related to the powerful neck muscles that attached across the bone in primitive humans. And in Neanderthals the profile of the occipital was influenced by the rather bulging occipital lobes of their brains, the significance of which is still debated—these lobes contain the visual cortex, for example. In modern humans, the parietals are heightened and lengthened, while the arch they make is narrower at the base but wider higher up. The paleoneurologist Emiliano Bruner investigated these aspects of ancient brain shape using geometric morphometrics. He confirmed earlier, more traditional, studies that blood vessel impressions on the inside of the parietals (reflecting blood supply to the parietal lobes) are altered in modern humans, forming a much more complex network.

So is there anything in the function of the parietal lobes that might explain their expansion in the modern human brain? They are involved in integrating sensory information, in processing data from different parts of the brain, and in social communication, all of which could be reflected in the behavioral changes recognized with the arrival of modern humans. The cognitive archaeologists Thomas Wynn and Frederick Coolidge argued that a key change in the modern human mind must have been the development of an episodic working memory. Memory in humans can be subdivided into declarative memories, such as basic facts and information, and procedural memories, such as strings of words or actions (such as making a tool or route finding). It is known from brain studies that these are separate modules, in the sense that brain damage may interfere with one but not the other, and brain imaging studies show they are controlled by different pathways. It is very likely that both of these types of memory are enhanced in the modern human brain, but there is one special and important type of declarative memory called episodic, personal, or autobiographical memory—a storylike reminiscence of an event, with its associated emotions. This can be used mentally to rerun past events, and, just as important, it can also rehearse future events—a sort of "inner-reality" time machine that can run backward or project possible scenarios forward, and which seems closely linked to concepts of self-awareness ("consciousness"). As we have seen already, archaeological evidence suggests that the reach of modern humans across the landscape in terms of food gathering, sourcing raw materials, and social networks increased during the Middle Stone Age and continued to increase during the Later Stone Age in Africa and contemporaneous industries outside of Africa, such as the Upper Paleolithic. Such developments could reflect the arrival of a modern kind of episodic memory. Moreover, the ability to conjure up vivid inner-reality narratives could also have been critical in the development of religious beliefs, since imaginary scenarios could be created as well as actual ones. Once people could foresee their own deaths, religious beliefs that provided reassurance about such events could perhaps have been selected for their value in promoting survival.

Experiments and observations suggest that the parietal lobes are indeed involved in episodic memory, but it is clear that they are not the only location implicated, since the recall of such memories involves a network of links between the frontal, parietal, and temporal lobes. Moreover, even episodic memory is not a single straightforward path. For example, some patients with selective parietal lobe damage can recall a particular event in detail from a general cue such as "your birthdays" (top-down recall), whereas others need a detailed cue such as a photo of a particular birthday cake (bottom-up recall) to remember one special event properly. But the lower parts of the parietal lobes are also implicated in another vital property of the modern human brain: inner speech. This is our inner voice that consciously and unconsciously guides so much of our thinking and decision making; in a sense it provides one of the most vital bits of software for the hardware of our brains. Indeed, there is evidence that an inability to create and use this program—for example, in people who were born deaf, mute, and blind, and who have been given little sensory stimulation from other humans—greatly limits higher brain functions. Even so, such severely impaired people, when given appropriate inputs from an early age, can develop and use their own codes of inner speech, for example, by recalling the symbols of sign language that they have been taught, instead of spoken words.

Stanley Ambrose, the champion of the impact of the Toba eruption on modern human evolution, also argued for the importance of memory and the development of particular parts of the brain in the success of modern humans. In his view, what was most important was the integration of working memory with prospective memory (dealing with near-future tasks) and constructive memory (mental time traveling), which are centered in the front and lower rear of the frontal lobes. Such links would have facilitated everything from the construction of composite artifacts to the development of the fullest levels of mind reading and social cooperation. In his view, archaic humans like the Neanderthals had developed the memory for short-term planning and the production of composite artifacts, but they lacked the full brain integration and hormonal systems that promoted the levels of trust and reciprocity essential for the much larger social networks of modern humans.

All of this shows how complex our brains are, and what a long way we still have to go to understand their workings in living humans, let alone ones who died 100,000 years ago. Unfortunately for Richard Klein's views of a significant cognitive event about 50,000 years ago, the heightening of the frontal and the expansion of the parietal lobes had apparently already occurred 100,000 years earlier, as shown by the shape of the early modern skulls from Omo Kibish and Herto. Overall, there is little evidence so far for any detectable changes in the modern human brain when Klein argues these should have occurred. Brain volume and EQ apparently increased fairly steadily in modern humans until the last 20,000 years, after which they seem to have declined somewhat, and similarly the trend in increasing cerebellum/cerebrum ratio seems to have altered only in the last 20,000 years or so.

Therefore, all we can say is that there is no obvious physical evidence for such a change in the workings of the human brain 50,000 years ago. Perhaps some genetic support will eventually emerge; there are claims that the gene DRD4, which when negatively mutated is linked with attention-deficit/hyperactivity disorder (ADHD), underwent changes around that time. DRD4 affects the efficacy of the neurotransmitter dopamine in the brain, and it has been suggested that a positive effect of such mutations would be to encourage novelty seeking and risk taking—perhaps important qualities for a migration out of Africa. John Parkington is one of several archaeologists and biologists who have argued that the fish oils obtained by early moderns when they began to seriously exploit marine resources would have boosted the brain power of early *Homo sapiens*—and there are further claims that omega-3 fatty acids would have conferred additional benefits in terms of health and longevity. But unfortunately, at the moment, such changes can only be inferred from indirect evidence such as the archaeological record, which itself can be interpreted in very different ways. Here we need to return to two of the key elements of modernity that might be decipherable from that archaeological evidence: the presence of symbolism and, by inference, complex language.

In chapters 5 and 6, we discussed some of the key behavioral "signatures" of modernity that are usually highlighted by archaeologists— things like figurative art and burials with grave goods. And we saw that there is not yet any strong evidence for figurative or clearly representational art anywhere before the European material dated at about 40,000 years. Equally, there is no evidence of symbolic burials older than the early modern examples known from Skhul and Qafzeh at about 100,000 years, even if older African sites like Herto are suggestive of the ritual treatment of human remains. However, the processing and use of red pigments in Africa does go back considerably farther, to beyond

250,000 years, at sites like Kapthurin and Olorgesailie in Kenya. The record is sporadic after this but emerges at Pinnacle Point in South Africa at about 160,000 years, and much more strongly at sites in North and South Africa from about 120,000 years. In particular, there is the rich material from Blombos Cave, South Africa, which includes about twenty engraved ocher fragments and slabs, dated to around 75,000 years ago and some which extend back to 100,000 years. These fragments seem to be generally accepted as symbolic in intent rather than accidental or utilitarian, but many of the earlier examples are only suggestive of symbolic meaning, rather than definitive.

The evidence seems much stronger in the case of tick shell beads, present at the known limits of the early modern human range at least 75,000 years ago, from cave sites in Morocco, Israel, and South Africa. But even here, the context in which they were being used becomes critical in deciding what level of symbolic meaning they carried. The archaeologist Paul Pettitt suggests an alternative way of looking at symbolic intent by moving away from judging an absolute (and contentious) presence/absence, and instead deconstructing different levels of symbolic meaning, in line with the Robin Dunbar stages of "mind reading" that we discussed in chapter 5. This deconstruction is valuable because it also allows an evolutionary sequence for symbolism to be considered, rather than just an on–off switch, where symbolism is either not there at all or fully developed, with no intermediates. Pettitt points out that symbols can only function as such in recent humans if the "writer" and "reader" are in accord over meaning, but in interpreting archaeological finds we tend to focus on the writer, without considering those who might receive the intended message. He cautions that unless the symbol is repeated at a number of different sites in a given time period, we should be cautious about how widespread was the behavior and how efficient it was at conveying its meaning.

The same symbol might have carried different messages for different individuals, and between different groups, while today we might consider only one of many possible intended meanings, and even that inferred meaning might be wrong. For example, for pigments and perforated shell beads applied to the body, there could have been different levels of use and symbolic meaning, from the simple to complex. The most basic use might be purely decorative and reflect a personal preference ("I wear red because I like red"). Or the message could be one of enhancement of the signal ("I wear red as I know you will read it as a sign of my strength or be impressed by it"). A third level might reflect status or group identity ("I wear red as I know you will recognize it as the regalia of our clan and infer from it that we are culturally the same"). A fourth and even more complex message might be "I wear red as, like you, I am a successful hunter and have killed an adult eland; it is my right to wear this color, and I therefore command respect from all." And, finally, the most complex, as part of an elaborate myth or cosmological belief, might be "I wear red only at this specific time of the year, marking when the ancestors created the land. This is a vital part of our beliefs, and by doing this I show that I am the bearer of this knowledge." Just considering the hypothetical examples above, at which of these levels would the tick shells and engraved ocher fragments from Blombos Cave have been functioning about 75,000 years ago? At one or several levels of complexity, and, if so, which ones?

We can probably rule out the simplest level because of the profusion of the shells and their consistency of selection and manufacture, and for the ocher, the engravings generally look carefully and specifically made in each case. But Blombos is an exceptionally rich example, and in other Middle Stone Age sites there may just be ocher crayons, with no engravings and no beads, so should we try and judge the level of symbolic intent at such sites? It is certainly possible that some of the earliest occurrences of red ocher in African sites were nonsymbolic, since ocher can also be used as a component of natural glues, as a preservative, or to tan animal hides. But equally, some may also have reflected a low level of symbolic intent, in terms of personal decoration and simple display. Indeed, the application of red ocher to human skin may have begun for purely practical reasons—as an insect repellent, say, or to hides as part of their production—but the red ocher was then favored for its attractive (and, later, meaningful) appearance. Personally I think the proliferation of shell beads and red ocher use along the length of Africa between 75,000 and 100,000 years ago must reflect an increasing intensity of symbolic exchanges both within and probably between early modern human groups. But perhaps the highest levels of symbolic meaning were still only nascent then.

Pettitt similarly deconstructs ancient burial practices into levels ranging from morbidity (an interest in the dead—shown even by chimpanzees) to mortuary caching (deposition of the body in certain places) to full burial in a special location, with ceremony, or accompanied by symbolic objects. In turn, he links these with Dunbar's levels of mind reading, so the simplest intentionality level (perhaps in apes and early hominins) might be "I believe that you are dead," followed by "I empathize that you are dead" (perhaps in early *Homo*), then "I know that you must be deposited in a specific place" (*heidelbergensis*, earliest moderns, and Neanderthals?), and finally "Because of your role, you must be disposed of in this way, by this method, at this place, as recognized by our social rules" (later modern humans and perhaps some Neanderthals?). It is possible that early humans like *heidelbergensis* were already treating their dead in some way, while Neanderthals were certainly caching and burying bodies with simple methods, with the possibility of more elaborate burials in some cases. But perhaps only modern humans carried out the most complex disposal practices for their dead.

What is puzzling here is that there are actually very few human burials known from the Middle Stone Age of Africa, and the best early examples come from the Middle Paleolithic of Israel about 100,000 years ago. In contrast, there are many late Neanderthal interments in western Eurasia, but we don't pick up the practice again in modern humans until about 40,000 years ago in North Africa, the Middle East, and then Europe. It's certainly possible that early moderns engaged in other methods of disposal than burial, just as people do today—in the open, on platforms, up in trees, or on ceremonial fires. This seems to be the case for some of the first moderns in Europe, too—the Aurignacians—whose bodily traces are mainly in the form of isolated and sometimes pierced human teeth rather than burials, suggesting that they preferred to carry traces of their enemies or their ancestors with them, rather than burying them.

Symbolism seems, then, to have been part of our African heritage, even if its recognition in the archaeological record is not a simple business. And what about language, which is assumed to have developed alongside symbolism? There are many different theories about the origin of human language, as we discussed in chapter 5, and it has been a source of great controversy since the time of Darwin—indeed, the venerable

Linguistic Society of Paris amended its constitution to ban any discussion of the origins of language in 1866! We know that human infants have a powerful built-in capacity to acquire and then use language, and they will readily learn whatever language, or even languages, they are exposed to. As Darwin recognized during his travels, there is no relation between types of societies and the complexity of their language. For example, linguists consider that English is one of the easier languages for a nonnative to learn compared with a host of other languages ranging from Hopi, Circassian (North Caucasus), and Kivunjo (Tanzania) to Arabic. Darwin favored an origin of human language through imitation and recognized parallels between human speech and birdsong. Various hypotheses have been proposed for the first stages of such imitation, whether of animal sounds, of natural sounds like wind or thunder, or of spontaneous exclamations such as pain or surprise, which gradually took on new meanings. A distinct set of hypotheses proposes that language arose through specific social needs, whether to avoid dangers, to facilitate cooperative hunting, or, as Leslie Aiello and Robin Dunbar proposed, to take over the social function of grooming as group sizes increased in early humans. And we saw that there are also models that postulate a much more sudden and serendipitous origin of complex language through genetic changes that fortuitously enhanced the relevant brain pathways.

From my perspective, I think that simple languages must already have existed in early human species, given the complexity of behavior that is apparent at sites like Boxgrove and Schöningen in Europe and Kapthurin in Kenya, and so Neanderthals would have inherited and built on the language or languages acquired from their ancestors. But in my view it was only with the growing complexity of early modern societies in Africa that sophisticated languages of the kind we speak today would have developed, through the need to communicate increasingly intricate and subtle messages. And by using the word *need* here I am not implying, of course, that the need created the desired result; what happened was that useful variations in human behavior and communication would have been enhanced through selection, and this could have included humanly driven cultural or sexual selection, favoring the best communicators. Our languages are not just for the here and now, as earlier ones probably were, since through them we can talk about the past and future, about abstract concepts and feelings and relationships, and about the virtual worlds that we can create in our minds. We humans are collectors and curators par excellence, storing and employing a rich vocabulary to name and describe the worlds we inhabit, both physical and virtual.

Finally, in this discussion of language, as we saw with the workings of our brains, the power of thinking itself would have been enhanced by an increasing richness of expression. So if complex language originated in Africa, could it have had one single origin—a protolanguage from which all tongues today have evolved? This idea of monogenesis, and with it the theoretical possibility that we can work backward from present-day languages to reconstruct at least something of the prototype, is certainly appealing to me. Several linguists have produced vocabularies of some of the hypothesized first words, and even something of the way in which they might have been used, but this is a highly controversial area, and one where glottogonists (*glottogony* comes from the Greek for "language origin") disagree strongly. For the moment, this is not an area in which we can draw firm conclusions, but the psychologist Quentin Atkinson has analyzed the number of phonemes (sound components) used in languages around the world and concluded that the global pattern closely mirrors that found in genes. Africa has the largest number and diversity of phonemes, and that number decreases as we move away from Africa. As Atkinson maintains, that would be consistent with an African origin for the languages of today, too.

In terms of innovation, we saw in chapter 1 that the apparently sudden florescence of the rich Upper Paleolithic societies of Europe seduced many in the last century to consider that this period marked the real arrival of fully modern humans, even if areas like the Middle East or Africa had been rehearsal grounds for the revolution that was to be finally expressed in the caves of France. But as we also saw, this Eurocentric viewpoint that the Cro-Magnons were the first "modern" people has been largely abandoned, although that is not to deny that something special did happen in the Upper Paleolithic of Europe. If Africa was actually at the forefront of Paleolithic innovations more than 40,000 years ago, why was that? As the anthropologist Rob Foley pointed out, the sheer size of Africa (one could easily fit China, India, and Europe into its surface area) and its position straddling the tropics certainly gave it advantages over any other area inhabited by early humans. The rapidity and repetition of climatic oscillations outside of Africa probably continually disrupted long-term adaptations by human populations in those regions.

Thus Neanderthals in Europe and the descendants of *Homo erectus* in northern China were constantly faced with sudden range contractions and the extinction of large parts of their populations every time temperatures sank rapidly, as they often did. And in the island regions of southeast Asia, where the descendants of *erectus*, and the Hobbit, and any similar relict populations lived, climate changes would have greatly disrupted connections between regions and populations, as sea levels rose and fell by one hundred meters or more. The local environments would also have been greatly affected by related changes in the monsoons and rainfall.

By comparison, in Africa, temperature and sea level changes were probably less damaging to its human inhabitants, and while there would certainly have been major changes in precipitation and environments, as we will discuss shortly, the continent nevertheless probably always had more people surviving there than any other region in ancient times. Given its larger human populations and its greater continuity of occupation, Africa probably always had more genetic and morphological variation than other parts of the inhabited world, giving greater opportunities for biological and behavioral innovations to both develop and be conserved. In this sense, it was perhaps more a matter of Africa being the place where early humans had the best chance of surviving, rather than being special in terms of a unique evolutionary pathway. This gives us an important clue to what eventually triggered our evolutionary success story.

As we saw in chapter 5, Africa shows the precocious appearance of features we associate with hunter-gatherers in recent times, such as planning ability, symbolic behavior, abstract thinking, marine exploitation, and enhanced innovations in technology. But, as we also saw, while such changes appear in some parts of Africa over 75,000 years ago, it is as though the candle glow of modernity was intermittent, repeatedly flickering on and off again. Most of the suite of modern features does not really take root strongly and consistently until much later, close to the time when humans began their final emergence from Africa about 55,000 years ago. Why was that? It is possible, of course, that the modern features were present in some African populations and not others, and as these groups moved around the landscape, their "modernity" seems to appear and disappear with their archaeological visibility. But I think another explanation is more likely, and this has to do with demography, the study of the size, structure, and distribution of populations through time and space, and the factors affecting population such as birth, death, and aging, migration patterns, and the environment. Valuable clues about the importance of demography can be found in the history of an island far away from Africa: Tasmania.

As sea levels have risen and fallen with the decline or growth of the Earth's ice caps, Tasmania has been intermittently joined to the mainland of Australia, and this was the case from about 43,000 to 14,000 years ago, allowing early Australians to reach there by about 40,000 years ago. Excavations in various caves show that the first Tasmanians adapted well to the cool, southerly conditions, with a variety of hunting and fishing weapons and tools and bone piercers that were probably used to manufacture the skin clothing and shelters that helped them survive the rigors of the last Ice Age. But from about 14,000 years ago, Tasmania was cast adrift from the mainland again by rising seas, and as the Earth warmed into the present interglacial, the landscape of the shrinking island changed to become more heavily forested. The original colonizers of Australia, including the ultimate ancestors of the Tasmanians, must certainly have made boats to reach the continent from southeast Asia, and those boats may have been made of bamboo. But if we asked some traditionally living aborigines in central Australia today to make us a seagoing boat (bamboo or otherwise), they would probably question our sanity, since this is no longer part of their lives and adaptations—in the absence of forests and large bodies of water. That situation seems to have applied in southern Australia and Tasmania when these regions separated at the end of the last Ice Age, meaning that their human populations lost contact with each other. While the mainland groups still had large areas of land with different environments available to them, as well as contact with each other, the Tasmanians now suffered from the previous loss of the knowledge needed to make seagoing craft. They were isolated on a shrinking island, with nowhere else to go, and this seems to have affected the pathways they then took.

Over the next 14,000 years, to judge from the archaeological record preserved in sites and the reports of the first Europeans to record their contact with aboriginal groups, the Tasmanians appear to have led an increasingly simplified life, forgoing apparently valuable skills and technologies, such as bone and hafted tools, nets and spears used to catch fish and small game, spear throwers and boomerangs, and anything but the simplest of skin clothing. Indeed, there were even reports that some groups had lost the ability to make fire at will, although this has been strongly disputed. Research suggests that these changes were the results of shrinking populations and loss of territories and resource bases, as well as the loss of contact with the mainland.

Anthropologists like Joe Henrich have argued that such changes interfered with the ability of populations to retain and pass on knowledge to each other and across the generations. For example, in a small population, an expedient short-term decision to, say, exploit seals on land for a few years rather than go fishing may have serious long-term effects if the knowledge needed to return to fishing had been compromised or even lost in the meantime. The resulting simplification of adaptations by the Tasmanians was the lowest-risk strategy to ensure survival in difficult times, but it was also accompanied by a loss of complex skills that would have been useful in the longer term. If there had been a subsequent return to glacial conditions and a lowering of sea level, conditions similar to those of the first Tasmanians could have been restored, but even then, unless they picked up lost skills from contact with the mainlanders, they would have needed to develop them from scratch, a process that could have taken many generations.

What happened in Tasmania may help to explain events in Africa more than 50,000 years earlier. As we saw, the evidence of modernity there is often disparate and discontinuous, like a flickering candle. What are we to make of archaeological sequences where typical Middle Stone Age assemblages are succeeded by apparently "advanced" Howiesons Poort artifacts, and then, in later deposits, typical Middle Stone Age material returns? Or what about the brief flowering of the Blombos people with their symbolic red ocher and variegated shell beads? We are perhaps misled by recent human history, where information storage in spoken, written, visual, or electronic form means that useful innovations are rarely lost, and the growth of "cultural" knowledge is incremental or even exponential. In the past, on the other hand, small populations would have been prone to population crashes or even extinction, or forced into relatively rapid movement or adaptation to survive, and this could have led to the regular loss of innovations that might have been useful in the longer term. Thus repeated "bottlenecking" did not just remove genes but also eradicated discoveries and inventions associated with the human populations concerned, and rapid environmental changes or population movements would have had the same effect. We can think, perhaps, of the people who crafted those beautiful wooden spears more than 300,000 years ago at Schöningen in Germany. Did that knowledge pass on continuously through to late Neanderthals over hundreds of millennia, or did a sudden temperature drop in northern Europe remove the spruce trees required to maintain that knowledge or even entirely wipe out those populations and their skills?

In nature, it is often argued, environmental challenges force evolutionary change, and we have the saying "necessity is the mother of invention." The price of a failure to innovate and survive is extinction, and the history of life on Earth is littered with extinctions—what can be seen as failed evolutionary experiments. However, the other side of the coin is that large, stable populations may have a greater ability to survive and to develop and maintain innovations, and I think that is actually the key to what must have been happening in Africa about 60,000 years ago. Research suggests that the optimal conditions for rapid cultural changes are those where there are large groups of interacting social "learners," and this is the case not only in humans but in our closest living relatives, the great apes. Studies of both orangutans and chimpanzees suggest that innovations in food acquisition and processing, including the basic use of tools, happen most often not when the environment is challenging or when groups have spare time on their hands, but when large social groups are in close proximity, allowing ideas to diffuse, and useful ones to thrive. Applying that conclusion to early humans such as Neanderthals

Our closest living relatives—chimpanzees—engage in basic tool
manufacture and use, in this case to crack palm oil nuts.

and modern humans would imply that the populations who progressed
the most culturally were not necessarily the most intelligent and skillful
(although those factors were important in the first place, of course), but
those who were able to network and pass on learning in large groups,
and to maintain those group sizes most consistently through time and
space. If modern humans had the edge over the Neanderthals and other
contemporary species in those respects, this may go a long way toward
explaining why our species began its successful expansion in and beyond
Africa, through our accelerating progress in cultural rather than physical evolution.

We discussed the evidence for larger social group sizes in modern
humans in chapters 5 and 6, along with the idea of a "release from proximity," the ability of members of our species to interact with each other
not just face-to-face at one time, as other animals do—and earlier human
species did—but also at a distance in both time and space through indirect symbolic communication. One view has been that such changes are

what precipitated the "Human Revolution" in Europe, with its extraordinary painted caves and continent-wide social networks, marked out by the movement of materials and innovations. But I think such changes originated in Africa, and as well as helping us survive there, such developments were the key to our ability to disperse and reach every habitable part of the world, and in doing so to displace or replace the other surviving human species. Building on some of Henrich's ideas, the geneticists Adam Powell and Mark Thomas and the archaeologist Stephen Shennan ran computer simulations for human groups at different population densities, allowing subpopulations to develop and exchange ideas with each other or not. The model showed that subpopulation densities could reach a critical point at which ideas and skills would suddenly accumulate. Density was thus important for developing new ideas, but migration between groups was also vital, to ensure that such ideas had a better chance to persist and thrive, rather than decay and perish. Thus for the survival and propagation of knowledge, it's not so much what you know but who you know that matters.

Powell and his colleagues also used genetic data to suggest that population size in Africa could have reached a critical threshold about 100,000 years ago, when population density and enhanced contact between groups could have allowed the rate of accumulation of innovations to overtake their loss, something probably rare in humans up to that time. Thus cultural change in the Middle Stone Age greatly accelerated, and the increased store of learning was beneficial to the survival of individuals and their groups. In turn this would have started a feedback mechanism, leading to a further increase in population density and contacts, and so on. What is interesting about this work is that it suggests that genetic continuity, large brains, and intelligence on their own will not ensure success for human groups; the survival of knowledge itself is also vital. This may go some way toward explaining why the Neanderthals with their large brains and evident intelligence could never make the leap forward that our species eventually managed.

They certainly made cultural breakthroughs, though—in burying their dead, producing blade tools, hafting weapons, and using pigments (predominantly darker ones than those in Africa). And, as already discussed, a recent study by João Zilhão and his colleagues of materials

Transmission in archaic humans?

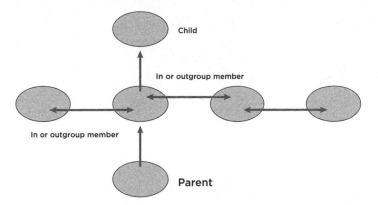

A diagram showing limited cultural transmission in archaic humans,
due to shorter life spans and smaller social networks.

from two southern Spanish sites, Cueva de los Aviones and Cueva Antón,
showed what they may have been doing with those pigments: they were
apparently mixing colors for cosmetics to apply to their bodies or faces.
Yellow, orange, red, and darker pigments were being mixed or painted in
or on seashells that had been imported from the coast, some of which
carried perforations (mostly natural, but carefully selected), making them
suitable as pendants. But despite these behavioral innovations, within
20,000 years or so the Neanderthals as a people were extinct.

It may be that, with the constant attrition of glacial climates, followed
by the arrival of modern humans, the Neanderthals were rarely, if ever,
able to maintain sufficient population densities to build on their achievements. And, as we saw, the process of cumulative innovation can go into
reverse, as it seems to have done in Tasmania and in Africa, even after the
behavioral features of moderns started to appear. There is also another way
of looking at the issue of why larger and denser populations might encourage innovation and change—and that is competition. Within human
groups there has to be a continual balancing act between cooperation and
competition for resources and mates. As I explain later in this chapter, the
development of religion may have provided an important means of maintaining that balance.

But as Darwin suggested, sexual selection could have been a powerful

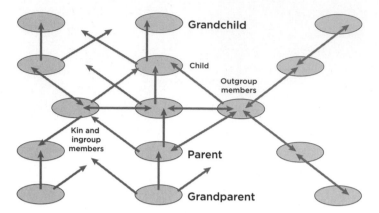

Transmission in modern humans?

A diagram showing much wider cultural transmission in
modern humans, due to extended life spans and much wider
social networks.

force within human as well as animal societies, an idea strongly champi-
oned by evolutionary scientists such as Helena Cronin and Geoffrey
Miller. They take the view that many modern human features such as
intelligence, creativity, and a way with words could have evolved not just
as survival tools in the face of a hostile environment but as courtship
tools, and through generations of mating preferences the genes that generated such behaviors were favored. Denser and larger human populations would also have engendered greater potential competition for
resources between neighboring groups. Perhaps they managed this competition through peaceful means, such as cooperation, in trading materials and partners, or sharing religious beliefs. But otherwise, conflict over
land, resources, or mates could have driven them into bouts of competitive innovation in behavior and technology, not only for gathering resources
but also for weapons. As we saw from chapter 6, it could also have driven
cultural and genetic changes favoring cooperative and even sacrificial
behaviors within the conflicting groups.

Saying that population increase was probably the critical factor in
conserving and building up behavioral novelties during the Middle
Stone Age of Africa still raises the question of what led to larger population densities and more extensive contacts between neighboring human

groups. There are many candidates for the agent in question, one of which is climate, which, as we already saw, can dramatically affect human population numbers. I will discuss this topic in more detail later in the current chapter. We also saw in chapter 6 that modern populations have achieved both better infant survival and greater longevity compared with earlier ones, and it may be that this process was already under way in Africa before 60,000 years ago, through better technology, provisioning, and mutual support. More grandparents would have meant more intergenerational knowledge transfer, and more support for mothers with dependent children, contributing to their survival. More grandparents would also have meant more kinfolk, providing wider kinship networks across time and space, valuable for exchanges of partners and goods, for alliances, and as an insurance policy against the time that your bit of territory suffered fire, famine, or drought—in other words, spreading the risk. And perhaps something like the institution of marriage and its associated extended family structures appeared at this time, cementing social ties between neighboring groups and catalyzing the growth of ceremonies, rituals, and symbolic exchanges.

In the same way, religion could have made a huge social impact. It was surely there, in the European Upper Paleolithic, with its depictions of what seem to be shamans and sculpted therianthropes, but I think evidence will emerge that it was also present in the later Middle Stone Age of Africa. In fact, some researchers hint that the richness of Blombos Cave may indicate that this was a sacred site, while claims have been made for the existence of a snake cult in a cave within the Tsodilo Hills of Botswana, based on a huge python-shaped rock within this Middle Stone Age site.

Something as important as the origin and growth of religious belief certainly warrants further discussion. This is an even more controversial area than the origins of language, with most scientists accepting that religion serves social needs and is deep-seated in humans—perhaps even with an inherited tendency, like the capacity to learn language. But a minority, echoing Karl Marx's words that "it is the opium of the people," see religion as a pathology—a crutch that people turn to when they are under extreme stress. As we saw, once the human brain had the potential for high levels of mind reading and for episodic memory, it was ripe for these to be co-opted for religious purposes (and indeed, brain scans suggest that similar cerebral pathways are used for religious thought). As I suggested in the discussion of episodic memory, did religion first provide a mechanism to allay possible neurosis about the future and about death, once we had the power of imagining these? Or, as Darwin believed, was it a natural consequence of human understanding of cause and effect—if there was an earthquake, or lightning struck, or the sun was eclipsed, or somebody died without an apparent cause, wouldn't the idea of supernatural agents such as spirits and gods have automatically followed?

Religion can certainly unite disparate and even geographically dispersed individuals to reinforce certain behaviors, and to give them a common purpose, but was the provision of emotional commitment or spiritual enforcement its original function and social benefit? Given the possibility of growing populations and more contact with other groups during the later Middle Stone Age, I think that the unifying effect of shared beliefs amid increasing social complexity would have been invaluable, providing the glue that bound people together, encouraging self-restraint and putting group needs ahead of their own. One can envisage a "successful" religion and the groups that followed it proliferating at the expense of other less successful, or nonbelieving, ones—and this competitive process continues even today. Indeed, as we saw in chapter 6, computer simulations have shown that in many situations of conflict between hunter-gatherer groups, beliefs that encourage self-sacrifice, sometimes including death on behalf of the group, can actually flourish culturally and genetically.

There is another potential benefit of religious belief, and that is the mnemonic (from the Greek for "memory") structures that religious myths can provide, structures that facilitate the storage and transmission of important information about the group, its history, and its environment. This is most powerfully demonstrated in the Dreaming creation myths shared by many native Australians. The Dreaming tells of the journeys of ancestral beings, animal or human, who molded the landscape and its individual features. These creator beings also passed down social rules and rituals for the maintenance of the land and the life it supports, and their journeys are marked by networks of Dreamtime tracks, joining sacred sites associated with these ancestors. One particularly widespread

myth is that of the Rainbow Serpent, a huge snake that lives in the deepest waterholes, and which was born from an even larger snake marked by
the Milky Way. It can manifest itself as a rainbow or can move through
rivers, shaping the landscapes and singing about the places it has made.
It may eat, drown, or infect those who displease it, while the righteous may
be blessed with powers to heal or make rain.

In the often hostile landscape of Australia, the transmission of Dreamtime myths through the generations must have saved many lives, since
the stories provide the equivalent of an outback sat-nav, leading people to
waterholes, food, shelter, and natural resources like stone and pigments.
Intricate legal, kinship, and territorial systems are also built into the local
variants of Dreamtime, so aborigines carry a virtual guide to life in their
heads, with segments of the stories recorded in paintings or carvings, and
sung or acted out at ceremonies marking the important stages of life
and death.

Dreamtime is a particularly all-encompassing creation myth, and
probably took many millennia to reach its present levels of sophistication, but less elaborate versions of such stories and mnemonic systems
probably existed in the Upper Paleolithic of Europe. Intricate carvings
on mammoth tusks from some of the Danubian Gravettian sites we discussed in chapters 5 and 6 may well be maps of rivers and the surrounding lands, while a 14,000-year-old engraved block from Abauntz Cave in
Spain was interpreted as showing mountains, rivers, and lakes, with
herds of ibex. Whether that is true or not, it seems likely that symbols,
ceremonies, and rituals acted like the Dreamtime stories in transmitting
the oral history of Paleolithic societies and reinforcing their rules of life
and their relationship to the landscape, even as far back as the Middle
Stone Age of Africa. It is sad for many reasons that the surviving Khoisan
(Bushman) peoples now occupy only a fraction of the range and environments that their ancestors did, to judge from the evidence of DNA, archaeology, widespread cave art, and linguistics. As we saw from the lessons of
Tasmania, their decline in territories and numbers must have affected
their cultural diversity, and we have thus lost the rich cosmological contexts
that must have lain behind their traditions of cave art, which stretched
back deep into the Later Stone Age. Similarly, the meanings of the enigmatic

Bradshaw paintings in the Kimberley region of northwestern Australia are lost to us, since they represent a now-vanished artistic tradition on that continent.

Returning to the Middle Stone Age, I want to look at the climatic record of that period, to see whether it holds clues to the growth of modern human populations and their innovative behaviors. When I was a student, the general view was that when Ice Ages hit Europe, Africa had a wet *pluvial period*, characterized by rises in lake levels. Similarly, when Europe enjoyed a warm interglacial, Africa suffered a dry interpluvial stage, with the spread of deserts. When I coauthored *African Exodus* thirty years later, in the 1990s, I proposed that a population crash in the severe global cold stage that lasted from about 130,000 to 200,000 years ago could have been the catalyst that drove the evolution of *Homo sapiens* in Africa. However, we now know that global climates are not read so simply, and climatic change in Africa often danced to a different rhythm from that of the major Ice Ages marked in expanding ice caps and falling sea levels.

In fact, different parts of Africa are affected by different factors. Studies of river and lake systems and of desert dust and pollen in offshore sediment cores show that varying conditions in the North Atlantic (for example, the chilling effects of Heinrich events) clearly influenced North and West Africa. However, the East was affected by the changing patterns of the monsoons off the Indian Ocean, while South Africa was influenced by conditions in the Southern Ocean, adjoining Antarctica. At times there were windows of opportunity for humans, with well-watered conditions in parts of Ethiopia 195,000 and 160,000 years ago, at Omo Kibish and Herto respectively, but the latter period was probably much more severe in the south, with sites like Pinnacle Point providing refugia near the relative stability of coastal resources. As we saw, during the warm interglacial about 120,000 years ago, the Sahara was "greened," with lakes and river systems, and the expansion of gallery forest and grasslands, favoring the spread of Aterian hunter-gatherers, with their distinctive tanged spear points, shell beads, and red ocher. But climatic data farther south suggest that central and southern Africa were generally more arid then, with many rivers and lakes suffering fluctuating levels, or even drying out completely. But after 75,000 years ago the situation reversed, with the Sahara turning to desert most of the time, while farther south much of Africa entered a cooler but more humid phase, with substantial rises in many of its lakes and river systems.

Human population sizes would thus have ebbed and flowed in Africa, sometimes extensive by ancient (but not modern) standards, and potentially in contact even across the full extent of what is now the Sahara, at other times pinned back in isolated refugia like Herto and Pinnacle Point. The complex climates of Africa may also explain why there seems to be no single center of origin for the earliest signals of behavioral modernity. Perhaps North Africa (and the Middle East?) led the way 120,000 years ago, but as conditions deteriorated, populations there shrank back or even became extinct, as favored environments rapidly vanished. Perhaps the torch of modernity was then kept alive farther south at sites like Blombos and Klasies River Mouth, as conditions favored that region for a while (give or take the interruption of events like the Toba eruption). Waves of population expansion and contraction could explain the brief but extensive florescence of the Still Bay culture with its rich symbolism, and the subsequent rise and fall of the Howiesons Poort with its innovative tiny hafted blades and engraved ostrich eggshells (recently described from Diepkloof rock shelter) more than 5,000 years later. And it is my guess (though we lack much data to support it) that East Africa became one of the next centers for behavioral evolution, about 60,000 years ago, as it was from there that modern humans (and their developing suite of modern behaviors) made their way out of Africa. My work with three geneticists on the recalibration of mtDNA evolution, discussed in the previous chapter, suggested that this was also the time of origin and first expansion of the L3 haplogroup, which gave rise to the M and N haplogroup families that characterize all of humanity outside of Africa.

So what can we say about the factors that led to the main modern human exodus from Africa about 55,000 years ago (assuming the earlier spread to regions like Israel was only a temporary range extension from Africa)? When considering movements of humans in ancient times, we usually take into account two factors—simply expressed as push and pull. The former is caused by negative factors, forcing groups to move as a result of a lack of resources, drought, or overpopulation. The latter is caused by positive factors such as the expansion of a favored environment or the promise of rich resources, inducing movement. And some ancient dispersals were no doubt entirely accidental and without motives, for example, where a group by chance tracked game into an entirely new region, or where a boat with people on board was carried to an unintended destination by unfavorable tides or winds.

Genetic data and physical proximity suggest that northeast Africa provided the immediate source area for the dispersal from Africa. There is evidence from Lake Naivasha in Kenya that East Africa was relatively well watered about 60,000 years ago, and data from both Antarctic and Indian Ocean cores suggest that the climate was relatively warm at that time, perhaps providing the right environment for population growth and another acceleration in innovation. There are many Middle Stone Age sites in Ethiopia, Kenya, and Tanzania that demonstrate human occupation around this time, but their study and dating are still largely in progress, and so it is difficult to relate them precisely to the time of the modern human dispersal from Africa. An example is the rich site of Magubike rock shelter in Tanzania, being excavated by Pamela Willoughby and her colleagues, where I am involved in the study of fossil human teeth. But there are published findings from one important site: a rock shelter near Lake Naivasha, Enkapune ya Muto (Twilight Cave). This lies close to sources of obsidian (a volcanic glass highly prized for making stone tools) at an elevation of 2,400 meters, with a rich record of Middle Stone Age and succeeding Later Stone Age occupation. The earliest levels of the latter industry show many innovative features such as specialized tools, red ocher use, and ostrich egg shell beads, dated to more than 46,000 years old.

In chapter 4 I discussed routes out of Africa and possible connections between North Africa, the Middle East, and Europe between 40,000 and 50,000 years ago. The most obvious pathway out of Africa would have been up the Nile Valley, through Sinai, and thence into the Levant (the region adjoining the eastern Mediterranean coast). Suggestions that another route could have existed at the other end of the Mediterranean—by boat across the Strait of Gibraltar—are appealing when one considers that the Strait was narrower during the last glacial, and there may even have been intervening islands at times. But good archaeological or fossil human evidence of an ancient connection is still elusive, with no sign of true

Neanderthals in North Africa, nor of moderns at an early date in southern Iberia. On the contrary, the region seems to have been one of the last outposts of the Neanderthals. It is possible that populations did cross from time to time but could not gain a foothold, but the fact that Mediterranean islands like Malta, the Balearics, Sardinia, and Cyprus do not seem to have been impacted by humans until much later also speaks against such mobility by early moderns and Neanderthals (although there are recent claims that African-looking handaxes were discovered on Crete).

But what about pathways farther east? We saw that there is evidence of the use of coastal resources in South Africa during the Middle Stone Age between 60,000 and 160,000 years ago, at sites like Pinnacle Point, Blombos, and Klasies River Mouth Caves, and the pattern is matched at sites scattered around the North and East African coasts, with a particularly interesting example at Abdur, on the Eritrean Red Sea, dated to the high sea level of the last interglacial, about 125,000 years ago. In the 1960s the American geographer Carl Sauer proposed that "the dispersal of early man took place most readily by following along the seashore. The coasts ahead presented familiar foods and habitats . . . Coastwise there was scarcely a barrier to the spread . . . through tropical and subtropical latitudes . . . The Indian Ocean, likeliest sea of earliest human occupation, exhibits to a large extent . . . an inviting articulation of shoreline, from Africa to the Sunda Islands [of southeast Asia]."

Some thirty years later the zoologist Jonathan Kingdon suggested that Middle Stone Age people left Africa through the Middle East and reached southeast Asia by 90,000 years ago, where they adapted to coastal living, including the development of a boat- or raft-building capability. This enabled them both to return westward to Africa and to move southward to Australasia. The anthropologists Marta Lahr and Robert Foley proposed in their Multiple Dispersals model that a more direct route from Africa to Arabia and farther east could have been taken before 50,000 years ago, perhaps using the coasts, and exiting by boat across the Bab-el-Mandeb strait at the southern end of the Red Sea. Geneticists such as Spencer Wells and Stephen Oppenheimer have also favored this route, arguing that only a few hundred people may have made the fateful crossing into Arabia, to found the populations of the rest of the world.

Personally I have never seen the necessity of invoking this complication in our African exodus, since if people were on the western coast of the Red Sea they would only have needed to travel north, around Sinai, and then south again down the eastern coast of the Red Sea. Continuing along the narrow ribbon of shoreline, to which they were already adapted, they could have progressed within a few millennia all the way to Indonesia at times of low sea level, and could have been spared the level of habitat disruption faced by inland populations during the rapid climatic fluctuations of the late Pleistocene. Population increase or depletion of resources would have driven the dispersal, and the development of watercraft could have followed from the need to traverse natural barriers such as dense mangrove forests or river estuaries, or to extend coastal foraging opportunities. These coastal populations could then have penetrated up river valleys into the interior, and by the end of their journey through southeast Asia, they would have been ready for the (probably fortuitous) first steps toward New Guinea and Australia (then joined as one enlarged continent).

The region of Arabia probably did form an important transit route for Out of Africa pioneers, but until recently it was largely terra incognita in terms of hard data, leading to much theorizing. Recently, however, several international teams have been trying to fill this void through fieldwork in the area. Archaeologists like Jeffrey Rose and Michael Petraglia have argued that it was not just a place that people passed through but an important locus for early humans in its own right. When environments were hyperarid in western Asia, it provided refuge with its ephemeral rivers and lakes, and its coastal margins, as these were exposed by falls in sea level. Despite the confounding lack of diagnostic fossil evidence, research by Simon Armitage and his colleagues at Jebel Faya (United Arab Emirates) provided important clues that early modern humans might even have dispersed from Africa, across Arabia as far as the Straits of Hormuz, by 120,000 years ago. This research augments the controversial idea that such modern populations could have migrated even farther across southern Asia, despite the conflicting genetic data that movements only occurred after 60,000 years. However, the fact that the artifacts at Jebel Faya look "African" and do not resemble those associated with the contemporaneous Skhul and Qafzeh people in the Levant signals yet more complexity in the exodus of modern humans from Africa. Could there have been separate early dispersals, one from East Africa into Arabia and another from North Africa into the Levant? And what was the fate of these different populations 100,000 years ago? Did they die out, did they survive in small pockets, did they perhaps interbreed with neighboring archaic peoples, or could they indeed have spread farther eastward?

After suffering colder and drier conditions, the climate in southern Asia improved with the temporary return of a strong summer monsoon about 57,000 years ago, and this may have helped the survival and migration of modern humans across India toward southeast Asia and Australasia at this time. However, signs of "modern" traits such as symbolism and complex technology are hardly apparent until after 45,000 years ago—a subject we will discuss later in this chapter in connection with Australia. And the Indian subcontinent does not have a single human fossil to record who was making its Middle Paleolithic tools; there is nothing between the archaic-looking braincase from the Narmada River gravels, which is probably over 300,000 years old, and fragmentary modern specimens from sites in Sri Lanka, dating from less than 40,000 years ago.

Work led by the archaeologist Michael Petraglia demonstrated there are Middle Paleolithic tools in Indian sites immediately before and some time after the widespread deposition of the Toba ash about 73,000 years ago, suggesting that whoever those people were, they may have been able to bounce back and repopulate after the apparent destruction wrought by Toba. Petraglia believes they were probably modern humans, perhaps descendants of groups like those known from Skhul and Qafzeh farther west, but if so, their mtDNA and Y-chromosome DNA have not survived today. This is because the M and N mtDNA haplogroups that exist throughout the region are probably younger than 60,000 years, while all Asian Y-DNA is even younger than this. So if these early Indians were modern humans (which is not yet demonstrable), they either became extinct or were largely replaced by later waves of modern human dispersal.

Recent DNA analyses showed just how remarkable was this spread of modern humanity across southern Asia. Study of some 55,000 *single-nucleotide polymorphisms* (*SNPs*—individual "spelling mistakes" in the genetic code) in about 2,000 people representing over seventy populations from right across Asia demonstrated that, despite clear physical

Paviland

Sunghir

Cro-Magnon
Abri Pataud

Mezmaiskaya

Neanderthal

Engis

Altamira
Cueva Morín
El Castillo

Dolni Věstonice

Teshik-Tash

Lagar Velho

Krapina

Oase

Gibraltar

Saccopastore

Shanidar

Dederiyeh

Irhoud

Taforalt

Haua Fteah

Üçağizli

Nazlet Khater

Taramsa

Adbur Reef

Singa

Herto

Adurha

Iwo Eleru

Eliye Springs

Omo-Kibish

Ishango

Guomde

Katanda

Kapthurin

Enkapune Ya Muto

Ngaloba
Mumba

Mumbwa

Kalambo

Twin Rivers

Cave of Hearths

Sterkfontein

Border Cave

Apollo 11

Wonderwerk

Florisbad

Blombos

Diepkloof

Sibudu

Klasies River

Die Kelders

Pinnacle Point

Map showing later human sites.

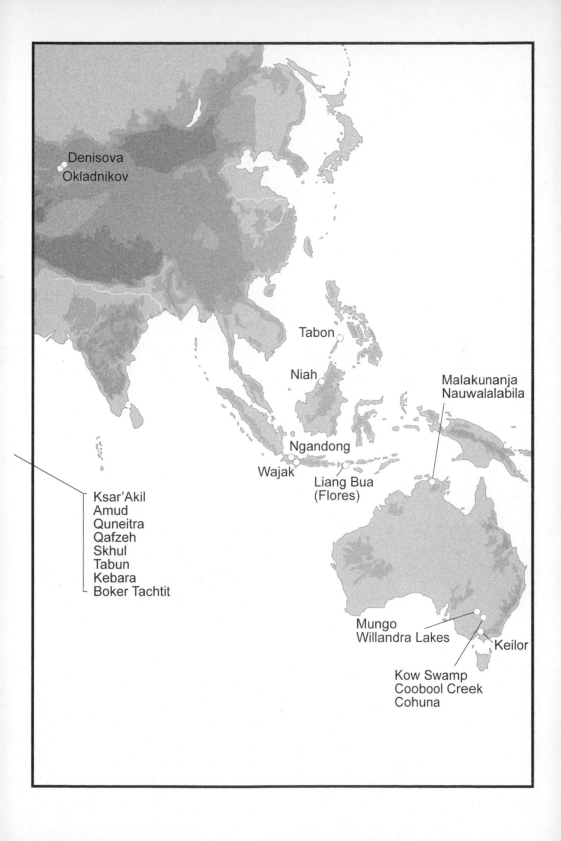

Denisova
Okladnikov

Tabon

Niah

Malakunanja
Nauwalalabila

Ngandong

Wajak   Liang Bua
                (Flores)

Ksar'Akil
Amud
Quneitra
Qafzeh
Skhul
Tabun
Kebara
Boker Tachtit

Mungo
Willandra Lakes

Keilor

Kow Swamp
Coobool Creek
Cohuna

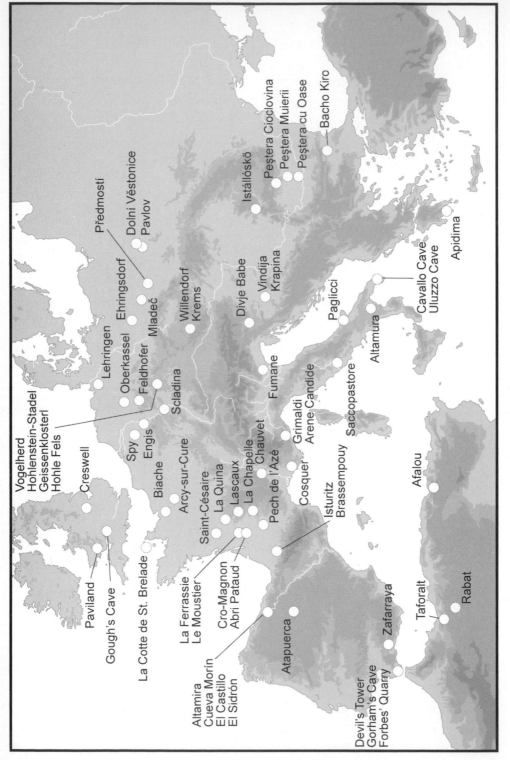

Map showing later human sites in Europe.

differences in appearance, skin color, and stature, the inhabitants of east and southeast Asia, including so-called Negrito peoples in the Philippines and Malaysia, are essentially one family of humanity (give or take some Denisovan DNA!) and derive from a single southern migration into the region. Genetic variation within the local populations decreases from south to north in east Asia, so subgroups moved north to found the less diverse populations of northern China, Korea, and Japan. However, the distinctiveness of central Asians suggests that they derived from a separate peopling of that region through the Eurasian Steppes. But groups like the aboriginal inhabitants of Hokkaido Island in Japan—the Ainu— were apparently not included in the analyses, so their origins in an even earlier dispersal of moderns, and a relationship to the first Americans, remains a possibility.

An intriguing and separate study investigated another SNP variant in Asian peoples called EDAR T1540C, and shows how DNA analyses are revolutionizing our understanding of human variation and how quirky some of our "racial" features may turn out to be. Many Asians are characterized by a hollowing on the back surfaces of their upper incisor teeth, called shoveling, since the inside surface resembles a tiny shovel— and a similar shape is also found in *Homo erectus* and Neanderthals. East Asians are also generally characterized by having straight and coarse black hair. The EDAR gene codes for proteins that are involved in the development of hair, teeth, and other derivatives of our skin and harmful mutations cause a condition called ectodermal dysplasia, where individuals may completely lack hair, nails, sweat glands, and normal teeth. The T1540C variant seems to be related to the production of both shovel-shaped incisors and coarse black hair, and is very common in East Asians, while it is virtually absent in Europeans and Africans. As yet it is unknown what produced the high frequency of this gene in East Asians. Was it chance or was it selection for one aspect—perhaps resistance to a skin disease, a particular kind of tooth strength, or thicker hair to protect against the cold, encourage the attentions of the opposite sex, or discourage the attentions of lice? Whatever the reason, it shows how features that have been considered very important in "racial" and even evolutionary studies may derive from quite unexpected factors—or no factors at all, other than as a by-product of something completely different!

An offshore core from the southern South China Sea records warm conditions and increased summer monsoons between 50,000 and 40,000 years ago, so early moderns who had arrived in the region by then were able to disperse northward under mild rather than glacial conditions—and this was probably the origin of the isolated Tianyuan modern human individual discussed in chapters 3 and 4. These East Asian pioneers were also able to move southward and reached Niah Cave in the Malaysian province of Sarawak, on the island of Borneo, by at least 45,000 years ago. This enormous cave, famous for its birds' nests which are used for the oriental soup, was partly excavated by Tom and Barbara Harrison more than fifty years ago and produced a wealth of archaeological material of different ages, but the most famous find was the "deep skull" of a modern human. This was controversially radiocarbon dated to about 40,000 years ago from associated charcoal, but many scientists refused to accept the validity of the determination, and it has taken until now for renewed excavations to show that the Harrisons essentially got it right. Moreover, the new work, led by the archaeologist Graeme Barker, showed that these early modern inhabitants of Niah had adapted quickly to the considerable demands of survival in tropical forests, since they were hunting many arboreal species and were processing local plants for their carbohydrates and others for dyes and pigments. However, their stone technology (and apparently their use of other local raw materials) was relatively simple—yet it clearly did the job very well. This reminds us of the important fact that being a modern human is as much about expediency and pragmatism as it is about harpoons, pendants, and cave art.

That is a crucial point to bear in mind as we trace the modern human diaspora toward Australia, because there we will search largely in vain in the earliest archaeological records for the markers of behavioral modernity that we have been discussing so far in this book. Evidence for complex stone or bone technology, structured sites, and symbolic behavior is generally lacking, although, to be fair to the earliest Australians, we already noted that the two 42,000-year-old Mungo fossils may represent the oldest red ocher burial and cremation so far discovered. To get there at all, these early Australians had island-hopped quite rapidly on boats or rafts across many stretches of open sea—the first long-distance seafarers of which we know—assuming, that is, that the ancestors of the Hobbits of

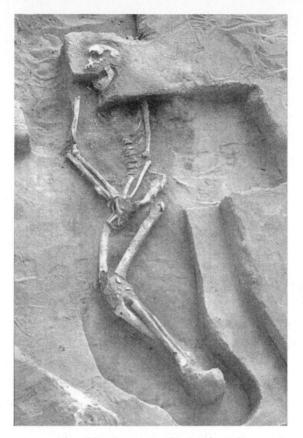

The oldest known red ocher burial:
Mungo 3 from Australia.

Flores had got to that isolated island by accidental rafting rather than purposeful navigation.

Yet the archaeological record suggests that the earliest Australians lived in small, highly mobile bands and exploited quite a narrow range of plant and animal resources, all of which could be obtained with simple technologies. The stone tools were much more like Middle Paleolithic ones than those we know from the Later Stone Age of Africa or the Upper Paleolithic of western Eurasia, and this led to the earlier suggestion that Australia was first colonized more than 60,000 years ago, before the development of the full suite of "modern" behaviors. However, from the latest dating analyses and genetic studies, it seems difficult to find evidence of colonization much beyond about 45,000 years. Although remembering the apparent simplification of life in Tasmania that followed isolation and decline in numbers, we can see how the same effect could have operated quite severely 50,000 years ago, as small numbers of pioneers survived the hazards of seafaring to reach New Guinea and Australia for the first time, and spread at low density around a vast and challenging continent. Yet in contrast, after 12,000 years ago, when native Tasmanians were apparently suffering a decline in numbers as the Earth warmed up, their mainland fellow Australians were increasing in population density in many regions.

As the archaeologists James O'Connell and Jim Allen pointed out, during the early part of the present interglacial, mainland aborigines underwent their own "Human Revolution," developing new ways of managing habitats, increasing technological complexity, and engaging in elaborate art and ornamentation, including of their bodies. At the same time there is evidence of a tremendous increase in the number of archaeological sites in regions like the Murray River, so this brings us back to the critical question of population density: was the benefit of increasing numbers the same as we postulated for the later Middle Stone Age of Africa, with greatly increased contacts, exchanges, and symbolic signaling between groups? And was the apparent lack of the signals of modernity in Australia before 12,000 years ago a parallel with the situation in Tasmania after that time, as shrinking and increasingly isolated bands of Tasmanians shed all but the essentials they needed for survival, while indisputably still representing "modern" humans?

This brings us back to the importance of climate, for, as O'Connell and Allen point out, the productivity of Australia was probably much reduced during cooler and more arid phases of the last 50,000 years, while the peak of warmth in the early part of our present interglacial, plus its relative stability, allowed mainland populations to thrive. And it remains possible that population densities (and hence the potential to express and accumulate signals of "modernity") fluctuated in Australia between 12,000 and 45,000 years ago. Perhaps the well-watered Willandra Lakes of 40,000 years ago allowed a temporary growth in numbers that catalyzed flashes of the expressions of "modernity" there—such as the exploitation of aquatic resources, red ocher use, and complexity in disposal of the dead.

This gives us an interesting thought-experiment to finish on, using that wonderful episodic memory of ours, before we move on to consider both the past and future of our species in the final chapter. We judge what it is to be human by the standards of our species, as we are the only surviving example available to study in detail. We are large-brained, fully bipedal, small-toothed, and good with technology (give or take an inability to work remote controls or mobile phones properly), and those are features we share to a greater or lesser extent with our extinct relatives in the genus *Homo*, such as *Homo erectus* and *Homo heidelbergensis*. But because people like the Neanderthals did not make it to the present day, we are the only surviving example of a "modern" human, so we look back at our evolution and assess "modernity" in terms of the development of "our" features: a high and rounded skull with a small brow ridge, large parietal lobes in the brain, narrow hips, and a bit of an obsession with religion, sex, and fashion (although not in my case, of course).

However, as I discuss in the last chapter, some of our "special features" and our differences from the Neanderthals might be there purely by chance, through the process of drift. And what if the Earth's climate had been subtly different over the last 200,000 years, or a volcano the size of Toba had gone off in Africa rather than Sumatra? What if those early modern humans had died out in Africa, and Neanderthals had managed to thrive in some part of southern Eurasia, building their population densities until they too fully—rather than partly—accomplished their own "Human Revolution"? Would they have eventually looked back at their success, at the benefits of the Eurasian environment, and the problems of surviving in Africa for their failed relatives, the ones with the weird foreheads? And would they have credited their big brains and occipital lobes, projecting noses, wide hips, ginger hair (sometimes), and the ancient use of black pigments for their evolutionary success, and used those to define a Neanderthal version of "modernity"? Even more remarkably, what if humanlike creatures had died out completely everywhere other than Flores, and the future of "humanity" then lay with the tiny Hobbit? Could it have survived and eventually spread from there to continue its own evolutionary story, and would its descendants have eventually pondered over an "Out of Flores" model?

We have traveled far and wide in this book, over millions of years and millions of square kilometers of the Earth's surface. Now it is time to take a hard and critical look at what we have discussed, to look at the past and future evolution of our species. Writing this book has been both a pleasure and a challenge for me, as I have had to revisit so many ideas of mine and those of colleagues. Along the way I hope that I have changed some of your perceptions of what it is to be human, just as I have changed mine.

9

The Past and Future Evolution
of Our Species

AS THE STUDY OF MODERN HUMAN ORIGINS HAS BLOSSOMED over the last forty years, I have been privileged to be involved in several lines of accumulating evidence—fossil, chronological, archaeological, and genetic—that our species had a recent African origin. And yet, as has become obvious, there are many loose ends to be tied up in our early history, and, more than that, there are still many fundamental questions that remain unanswered. We have partial answers to the "when" and "where" of our origins, but still precious little about the "whys." The basic reshaping of skull form to the "modern" pattern had occurred in Africa by 150,000 years ago, but the underlying factors remain largely unknown. Some may well have related to changes in the underlying brain shape, and others to quite different functional factors connected to our jaws, teeth, head balance, or vocal or respiratory tracts. However, there is an alternative and more mundane explanation for many of our "special" cranial features, which suggests that they were not really special at all—they were caused by genetic drift, essentially by chance.

I collaborated in research with two evolutionary anthropologists, Tim Weaver and Charles Roseman, and I can't do better than adapt a summary of our paper.

We used a variety of statistical tests from quantitative- and population-genetic theory to show that genetic drift can explain cranial differences between Neanderthals and modern humans. The tests were based on thirty-seven standard cranial measurements from about 2,500 modern humans from thirty populations, and twenty Neanderthal fossils. As a further test we compared our results with those for a genetic dataset consisting of 377 microsatellites typed from 1,056 modern humans from fifty-two populations. We concluded that rather than requiring special adaptive accounts, Neanderthal and modern human crania may simply represent two outcomes from a vast space of random evolutionary possibilities.

In fact the differences between modern human crania are fairly marked, despite our close genetic relationship, which is why forensic tests of regionality work so well, and as we saw, much of this could be the result of an accumulation of differences through drift, as modern humans spread out in small numbers. The variations we all show when compared with Neanderthals are even more marked, and as Erik Trinkaus pointed out, we diverged farther away from a potential ancestor like *heidelbergensis* than the Neanderthals did. Nevertheless, many of the contrasts may be the result of the same process writ large, the consequences of a geographic separation from our Neanderthal cousins about 400,000 years ago, after which we drifted—rather than were driven—apart. And it is possible, as Weaver suggested, that primates and earlier humans were much more constrained by natural selection to keep to a given skull form, whereas we (and perhaps, to an extent, Neanderthals) were able to escape those constraints through our cultural adaptations, allowing our cranial shapes to drift more randomly.

There does seem to be something different about modern human "cultures" when compared with those that came before. Ours seem to vary much more, and over much shorter periods, than those of the Middle Paleolithic, and even more so when measured against the limited variety of the Lower Paleolithic, which has been likened to 2 million years of boredom. There are probably many different reasons behind our cultural diversity; one is the sheer geographic range of modern humans and the variety of environments in which we live, and to which our ways of life are adapted. And yet the Upper Paleolithic people of Europe who lived in one region during one, albeit fluctuating, cold stage also showed great varieties and changes from one "culture" to another every few millennia. The anthropologists Robert Boyd and Peter Richerson argue that a key feature of *Homo sapiens* is our ability to imitate, and to learn from, each other—a trait that is there in the youngest of our children. And human societies give us the arena in which to rehearse, test, and modify what we learn through dialogue with our peers; as we saw, the denser the networks of connectivity and the wider the reach of those networks, the better. Boyd and Richerson also argue that with the rapid changes of environment suffered by humans in the past through climate change (or equally, I would add, the rapid dispersal of small numbers of modern humans to new and unfamiliar landscapes), there was no benefit in individuals separately trying to innovate their way to survival, and certainly no use waiting for evolutionary adaptations to come along—there simply would not be time.

Surprisingly, the best strategy for the average person in a variable environment might be to look around and rely on imitation, rather than individual learning. Some people might discover ways of coping with the changes, and, by imitation, this would provide a shortcut to success for the many. Moreover, the more copiers there were, the more chance that one of them might not copy accurately and by chance produce an accidental improvement on the original. If individual innovation was rare, progress would be slow—hence, perhaps, the 2 million years of boredom of the Lower Paleolithic. But if there were more innovators, the process could produce rapid cultural adaptations very quickly compared with individuals innovating in isolation, or with biological evolution. Thus through imitation and peer-group feedback, populations could adapt well beyond the abilities of an isolated genius, whose ideas might never get beyond his or her cave or might be lost through a sudden death. Boyd and Richerson contend that evolution tracked these developing tendencies for social learning. Thus selection would have favored minds that, rather than always attempting to invent something new (a risky and wasteful strategy overall), would have tended to conform to the majority behavior.

Evolution would have particularly favored a tendency to imitate high-status individuals in the group, or successful peers, and we can see how this could have led to group identities being symbolically displayed, and to the growth of fads and fashions, useful or not. *Style* is used both by archaeologists to label artifacts in past cultures and by people in the fashion industry today, and it may well explain a lot of the puzzling variations that rapidly come and go in the Upper Paleolithic record. Boyd and Richerson also looked at recent hunter-gatherer groups and how they have managed to occupy such an astonishing range of habitats. When we look at such groups in, say, the rain forests of Brazil or New Guinea, what is astonishing is their diversity there too, and the way they are often split into groups with mutually unintelligible languages, distinct traditions, and symbolically differentiated identities. Imitation-based cultures give these groups the capacity to behave somewhat differently from each other, and therefore the potential to track and exploit separate parts of their habitat—in a sense, to behave like different species in the way that they partition their use of the environment.

So this is the other side of cultural evolution, and there must always have been a tension in modern humans between the need to maintain and build on innovations, which, as we saw, requires high densities and connections between people (fusion), and the tendency of populations to go their own way and become increasingly culturally subdivided (fission). These alternative trajectories would have different genetic consequences too, given the need to maintain a healthy gene pool through partner exchanges with the neighbors (unless these were achieved through raids and kidnaps, which do seem to be surprisingly common, considering the "peaceful" image we may have for hunter-gatherers).

My fellow Neanderthal researcher Jean-Jacques Hublin worked with the evolutionary anthropologist Luke Premo to model the effect of culture in directing gene flow in humans—modern ones, certainly, but perhaps also the Neanderthals. As we saw in chapter 7, one of the striking things about both Neanderthals and *Homo sapiens* is their low genetic diversity compared with our primate relatives, and the implication that this must derive from low population numbers (bottlenecks) in our respective evolutionary pasts. But Hublin and Premo's modeling suggested that long-term culturally channeled migration—that is, between population subdivisions sharing cultural values—would have acted to suppress the effective population size of Pleistocene humans, just as bottlenecks do. We already saw that some modern human populations in Africa such as the Bushman and central African "pygmies" have deep and seemingly separate roots back to the time of the Middle Stone Age for some genetic markers, and this suggests that 100,000 years ago Africa may have comprised a collection of separate subgroups who predominantly exchanged genes internally, rather than across a single continuous population—perhaps because of geographic isolation, but perhaps also because they were subdivided by cultural distinctions and hence genetically relatively "inbred." This is something to which I will return later in this chapter, after further consideration of hybridization between early *Homo sapiens* and other human species such as the Neanderthals and *Homo erectus*.

As discussed in chapter 1, my view of *Homo sapiens* and the Neanderthals as representing distinct species has never led me to say that interbreeding between them was impossible. My old friend Erik Trinkaus is convinced that this would actually have been the norm, rather than the exception, as modern humans spread from Africa; as he succinctly puts it, "sex happens." He sees widespread evidence for it from quirky anatomical features in early modern fossils ranging from Portugal to China which, according to him, most likely derived from such intermixture. While certainly not ruling this out everywhere, I see the evidence differently and consider that many of these features were probably part of the normal variation of early *Homo sapiens*, rather than necessarily emanating from interbreeding with our archaic relatives.

Hybridization is also known as *reticulation*—the netlike pattern produced by gene flow between evolutionarily separate lineages—and in mammals it is apparent at several levels of the taxonomic hierarchy. Most often it occurs between closely related species (for example, monkeys within the widespread genus *Cercopithecus*), but sometimes it can occur between different genera (such as African and Indian elephants), and occasionally between much more distantly related forms (for example, through artificial insemination, between the Asian camel and the South American guanaco). Sometimes hybridization has only been detected through genetic studies of seemingly "pure" species, revealing that they had exchanged genes at some time in their evolutionary past, and its impact on the populations concerned may range from damaging (for example, where the hybrids are deformed or infertile) through to advantageous (for example, the rarer outcome of "hybrid vigor," or heterosis). As these examples show, they undermine, or at least modify, the concept of a biological species as genetically

self-contained and watertight from extraneous gene flow, and they emphasize the fluidity of species concepts, which are, after all, humanly created
approximations of reality in the natural world.

Many studies have examined the human genome for signs of interbreeding with our archaic relatives, and as we saw in chapter 7 this has even been
investigated through the DNA of our head lice. Up to now, the big picture,
from our autosomal, mitochondrial, and Y-chromosome DNA, has generally lacked signs of introgression from other human species, although
scientists such as John Relethford, Vinayak Eswaran, Henry Harpending,
and Alan Templeton have argued that indications were indeed there. Short
branches in our gene trees, particularly in Y-DNA and mtDNA, pointed to
a simple, recent African origin, and simulations from mtDNA data of the
level of possible Neanderthal and Cro-Magnon admixture suggested that
it was either zero or very close to zero. However, despite the fact that mtDNA
and Y-DNA provide such clear genealogical signals, they constitute only
about 1 percent of our total DNA, and signs of hybridization were clearly
lurking in the rest of our genome. Inconsistencies and exceptions to the
simple patterns shown in Y-DNA and mtDNA now suggest a more complex
evolutionary history for our species. This is an area of exciting current
research, particularly with the arrival of Neanderthal and Denisovan
genomes for comparison with the growing number of modern ones from
all over the world.

A recent example of such work is the study by Michael Hammer and
his colleagues. They used DNA sequences from sixty-one noncoding
regions in the autosomal genome of three sub-Saharan populations (the
Mandenka of West Africa, Biaka pygmies from central Africa, and San
Bushman from southern Africa) to test for signs of archaic admixture in
Africa itself. They concluded that African populations also contain about
2 percent of ancient genetic material, and this was input some 35,000
years ago, not from Neanderthals or Denisovans but from an unknown
archaic population within Africa itself, which might have been separate
from the modern human lineage for some 700,000 years. Three separate
regions of DNA showed deep divergences and one seemed to have intro-
gressed from an ancient and now-extinct population in central Africa.
Such a deep time scale could suggest that early groups of *heidelbergensis*
survived in Africa alongside the evolving lineage of modern humans.

So let us briefly review what we know from fossils about human species about 200,000 years ago. In East Africa, we see the first known signs of the modern human pattern of a high rounded skull, while in Europe, Neanderthal evolution was well under way. We have virtually no data on who was living in western Asia at that time, while the more ancient Narmada skull from India suggests that a descendant of *Homo erectus* or *heidelbergensis* might have been living there. China seems to have been occupied by descendants of *Homo erectus* or *heidelbergensis* too, but because of uncertain dating we cannot be sure whether the Ngandong fossils indicate that *erectus* was still surviving in Java. But *Homo floresiensis* was presumably well established in its long and lonely residence of Flores. That seems to represent a relatively clear picture in Africa and Europe, if imperfect elsewhere. But let's home in on the African record in a bit more detail, and with a more critical eye than I have employed so far.

As we saw, Omo 1 at about 195,000 years and the Herto crania at about 160,000 years seem to establish an early modern presence in Ethiopia, and the Guomde fossils from Kenya might even take that presence back to 250,000 years ago. The fragmentary but more primitive-looking Florisbad skull from South Africa is about that age, but there are other fossils from East Africa that do not fit the pattern so well. These include Omo 2, with its more primitive braincase shape and angled back, and a still undated archaic skull from Eliye Springs in Kenya. In my view the Laetoli H.18 (Ngaloba) skull from Tanzania, which is thought to be only about 140,000 years old, is not a modern human, and I'm doubtful whether the similarly dated but pathologically deformed skull from Singa in Sudan is fully modern either. When we move across the Sahara to Jebel Irhoud, I doubt the modernity of the fossils from there too, even if the synchrotron did show that the child's jaw displays a slow and modern growth pattern in its developing teeth. Those fossils are currently dated to about 160,000 years now, although there are indications that they might be even older than that.

In my early analyses of the cranial shape of these African fossils, I considered them separately, which planted the seeds of a view that specimens like Omo 1 and Jebel Irhoud (if ancient enough) might be part of our ancestry. However, as I became more confident about the nature of the African sample and its dating, I started to lump all the late archaic

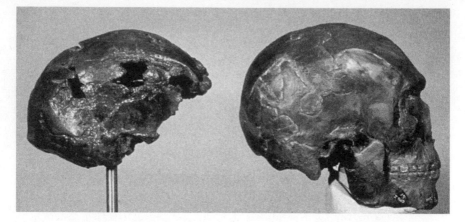

African variation 195,000 years ago: Omo Kibish 2 (*left*) and 1.

specimens together in my comparisons with samples from elsewhere, such as Skhul and Qafzeh, the Cro-Magnons, and the Neanderthals. I increasingly came to the realization that this was a mistake, because it obscured the considerable internal variation in that African sample.

Now, a similar conclusion about its variability was reached by the anthropologist Philipp Gunz and his colleagues, using geometric morphometrics. They measured the shape of the upper part of the skull in a sample of fossils ranging from *erectus* to Cro-Magnons and in a global collection of modern humans. They found, as I did, that African specimens like Irhoud, Omo 2, and Ngaloba occupied the middle ground between archaics like Neanderthals and *heidelbergensis* and the moderns, including Cro-Magnons and Qafzeh. But the African fossils showed more shape variation from each other than any equivalent group of fossils in the entire sample, whether modern, Neanderthal, *heidelbergensis*, or *erectus*. Might this be telling us something important about modern human origins in Africa? I have argued that *heidelbergensis* in Africa, represented by fossils like Bodo (Ethiopia), Elandsfontein (South Africa), and Broken Hill (Zambia), probably evolved into modern humans in some part of the continent, and I have wavered over whether this transition was a gradual or a rapid event. The general consensus has been that the transition was gradual, and that dating fossils like Broken Hill to about 500,000 years ago allowed plenty of time for the required major evolutionary changes in skull form (and probably body form, brain, and behavior) to occur.

But even the existing fossil record can still surprise us. Fifty years after Darwin published his cautious prediction that Africa would turn out to be the continent of our origin, the Broken Hill skull was fortuitously discovered, and so began the process of proving he was right. But the skull was not immediately recognized for what it was, and might easily have been lost to science. On 17 June 1921, miners quarrying a small hill full of metal ore near the town of Broken Hill (now Kabwe), in what was then Northern Rhodesia (now Zambia), uncovered a skull coated in brown sediment. Its huge staring eye sockets apparently scared them so much that they all ran off. Their supervisor, a Swiss miner called Tom Zwigelaar, was somewhat braver and got someone to photograph him at the site of discovery holding the relic.

The Broken Hill cranium ("Rhodesian Man") was the first important human fossil ever found in Africa, and even now it is one of the most impressive. It resides in a metal safe outside my room at the Natural History Museum in London and is one of the treasures of the Palaeontology Department—shiny brown and beautifully preserved, with massive brow ridges glowering over those empty eye sockets. It was first put into its own species—*Homo rhodesiensis*—by Arthur Smith Woodward of the museum in 1921 and has been named and renamed ever since. In 1930 it was described by the Czech-American anthropologist Aleš Hrdlička as "a comet of man's prehistory" because of the difficulty of deciphering its age and affinities. Despite its completeness and apparent primitiveness, its exact place in human evolution still remains unclear, because it has never been properly dated. When I was a student, it was still being used as evidence that Africa was a backwater in human evolution, because such a primitive specimen was living there only 50,000 years ago, while much more advanced humans had evolved in Europe and Asia. Now, as I said, it is usually placed with fossils like Bodo and Elandsfontein as representing the African component of our ancestral species *Homo heidelbergensis* from about half a million years ago.

As a small boy, visiting with my parents, I can vividly remember seeing the Broken Hill skull (or rather a plaster cast of it) on display at the Natural History Museum and being intrigued by its primitiveness and mystery. Ever since, I have nurtured the hope that I could help to place it definitively in the story of human history, either as an ancestor or as a distinct species that had died out without contributing to our evolution. I studied it for my Ph.D. in 1971, and I have regularly included it in my analyses of fossils, making it a central part of my concept of *Homo heidelbergensis* as a species that represented the last common ancestor of modern humans and Neanderthals. But without knowing how old it was, its precise place in human evolution has remained elusive and seemingly beyond reach, given the complete destruction of the site from which it came. But at last this is proving possible—and the results are surprising me as much as anyone else. For at least fourteen years prior to the discovery of the skull, miners had been digging through a fifteen-meter-tall column of fossilized bones, and because these were heavily impregnated with mineral ores, they were throwing them all into the smelter—I'd rather not think about what might have been lost!

After its discovery, other fossil human and animal remains and artifacts were recovered in the site and around the locality by people like Aleš Hrdlička and Louis Leakey, including from the mining dumps and the miners' huts, but only two human bones were found close to the skull, and at the same time. These were a long and straight shinbone (tibia) and the middle part of a thighbone. The latter find had a particularly interesting history. It was found by a Mrs. Whittington, who happened to be visiting her sister, whose husband worked at the mine. She was obviously an adventurous woman and was lowered down on a rope to collect it, but it was then virtually forgotten until Desmond Clark negotiated its transfer from a Rhodesian museum to London in 1963. Two other intriguing nonhuman finds also proved to be important in unraveling the lost history of the Broken Hill skull. One was a thin and mineralized yellowy brown silty deposit, which the miners collected because they mistakenly thought it was the mummified skin of Rhodesian Man. The other was a mass of tiny bones found around the skull and even cemented inside it. Originally thought to be those of bats, these actually represent the bones, jaws, and teeth of various small mammals, and they provide important information on the age of the skull and on where it originally lay in the cave.

The skull itself shows a strange combination of features. On the one hand, the brain size is only just below the modern average, but on the other, the face is big and the braincase shape decidedly primitive—long and low with enormous brows (*monstrous* was the word that Hrdlička used), with an angulated back to the skull and a transverse bony ridge reminiscent of *Homo erectus*. The cheekbones are not hollowed out as in modern humans (although a second upper jaw fragment found elsewhere in the site does show this feature), and the teeth are riddled with disease to an extent unusual for an early human: many are decayed and some of their roots are abscessed.

There are several other curious features, including a small and nearly circular hole in the left side of the braincase. Over the years, this has been suggested to be from a spear point, a lion's canine, or even primitive surgery. But not long after I joined the Natural History Museum, I learned of an entirely new idea. A British newspaper was serializing a book called *Secrets of the Lost Races* and requested permission to print a picture of the Broken Hill skull. When I asked what caption would accompany the illustration, I was told that it would say that this was the skull of a Neanderthal shot by an alien's bullet 100,000 years ago! I pointed out that the fossil wasn't really that of a Neanderthal, that it was probably much older than 100,000 years, and that a bullet hole would probably have been accompanied by radiating cracks; I also asked, what self-respecting alien would be using something as primitive as bullets? Nevertheless, it was agreed that the newspaper could have its photo if it included the statement that recent research suggested the hole showed signs of healing and was probably caused by disease emanating from within the braincase. Of course such scientific data didn't suit the paper's agenda, and it included a drawing of the skull instead, leading me to suffer several frustrating weeks as members of the public telephoned, wrote, or even turned up unannounced at the museum asking to see "the Neanderthal shot by the spaceman"!

The tibia found with the skull represents a tall individual of about 180 centimeters, but for a presumed male *heidelbergensis*, he was probably not that heavy at around seventy-five kilograms, compared with estimates for the fossils from Boxgrove (tibia) and Bodo (skull) of over ninety kilograms. So Rhodesian Man probably had the tall and relatively slender build we might expect for an inhabitant of the drier tropics today, although more robust and muscular. The tools from the site are a varied bunch, and none can be closely associated with the skull except for a round stone found with the femur fragment and which, along with others from the site, has been interpreted as a projectile, a pounding stone, or even a bola (a hunting or herding weapon formerly used in South America, consisting of balls connected by string or rope and which are thrown to entangle the legs of the target animal). Other artifacts included flakes, scrapers, and even some possible bone tools, but none of these look very ancient, suggesting a maximum age of perhaps 300,000 years. While some of the animal bones found around the site may have been those of prey, no studies so far have found convincing evidence of butchery, and given the complete destruction of the site, it is impossible to tell whether the bones and stone tools come from inhabitants of the cave(s) in the mined hill or found their way in through some other means.

I have been working on trying to date the Broken Hill skull more precisely for about fifteen years, with a number of collaborating scientists and even mineralogist colleagues at the Natural History Museum. The main methods we have been using (see chapter 2) are ESR (electron spin resonance) on a tooth fragment from the skull and uranium-series dating from various bones and sediment from the site. It would normally take great care and courage to remove an enamel fragment from one of the precious teeth of the Broken Hill skull, but a previous accident worked in our favor in this case. Some unknown member of staff or a visitor had accidentally knocked the corner from one of the molars and, rather than reporting it, had simply glued the piece back on! When one of my eagle-eyed colleagues, Lorraine Cornish, spotted this, she dissolved the glue and we had the perfect fragment of enamel to date. But one of the unknowns in ESR dating is the degree of radiation received by the enamel fragment since it was buried, and this has to be reconstructed from site data, which in the case of Broken Hill was severely lacking, with the complete destruction of the original location by mining. However, chunks of sediment and bone breccia were saved from the mine, in some cases because they also contained interesting minerals, and others were collected after the discovery of the skull, so these could be measured to help reconstruct the burial environment.

One of the worst-case scenarios in ESR dating is underwater burial, since water interferes with the accumulation of the ESR signal. There was plenty of evidence from the mine records that the level where the skull had been found had to have water regularly pumped out as it actually lay below the existing water table. However, two other clues I already mentioned became critical here. First, the "skin" that the miners thought they had found was actually layered sediment impregnated with minerals, which must have been laid down relatively horizontally and could not have formed under water. Records made at the time of discovery stated that it lay near the skull and tibia, at a steep angle, suggesting that it had fallen from higher up. Second, we know that the skull was covered in, and even contained, many bones of small animals such as shrews, and the mine records clearly document layers of small mammal bones at a much higher level, far above the place where the skull was actually found. So it seems likely that quarrying at the base of the sediments led them to constantly collapse down, and the skull was almost certainly derived from the higher levels of the site, above the water level.

Now, when we factor everything together, the ESR signal in its tooth enamel suggests that the skull is actually between 200,000 and 300,000 years old. And two other age estimates from the femur fragment and the so-called skin suggest the real age could be closer to 200,000 rather than 300,000 years. It is possible that the skin accumulated above the level of the skull and femur before they all collapsed down, so it could be somewhat younger than them, but certainly there is nothing here to indicate that this assemblage is anything like 500,000 years old. Such a surprisingly young age is not contradicted by the artifacts known from the site, which have early Middle Stone Age affinities, nor from studies by the paleontologists Margaret Avery and Christiane Denys of the small mammal accumulations closely associated with the skull, which in species represented match those known from African sites like Twin Rivers, dated in the range 200,000 to 300,000 years old. If the Broken Hill skull, one of the best-preserved relics of *Homo heidelbergensis*, is actually less than 300,000 years old, what does this mean for our models of human evolution and for the origin of our species?

The result does have important implications for our reconstructions of recent human evolution because, as I explained, the Broken Hill fossil has been a cornerstone of the assumed gradual evolutionary sequence from archaic to modern humans in Africa. Dating Broken Hill to about 500,000 years placed it some 300,000 years older than the first known modern humans, allowing plenty of time for the necessary changes to happen. But the new dating makes Broken Hill only somewhat older than Omo 1 at about 195,000 years, and perhaps close in age to the more modern-looking Florisbad and Guomde fossils. This would imply either that there was a very rapid evolutionary transition to the earliest modern humans about 250,000 years ago, or that Africa contained great variation in its human populations at that time. Could that variation have even extended to the coexistence of different human species? We already discussed the puzzling variation between the two apparently contemporary Omo Kibish fossils 1 and 2, with skull 1 looking decidedly modern and skull 2 having a more primitive braincase, with an angled back, and we mentioned the rather archaic-looking rear of the Herto adult skull. Moreover, there are other primitive-looking fossils in Africa (such as those from Ngaloba and Eyasi in Tanzania) that overlap the dates we currently have for the oldest modern-looking humans, and I will discuss a particularly striking example next. All of this means that I am reconsidering many of my previous views on the origin of our species in Africa, and I now think we need to talk about origins, rather than a single point of origin.

I pointed out in the previous chapter that the nature of the manufacturers of Paleolithic tools from many parts of Africa remains completely unknown, since there are no associated fossils. This is especially true of artifacts from West Africa, where the oldest known fossil, from the Iwo Eleru rock shelter in Nigeria, is thought to be less than 15,000 years old. This poorly preserved skeleton was excavated from basal sediments at Iwo Eleru in 1965 by the archaeologist Thurstan Shaw and his team and was associated with Later Stone Age tools. That latter fact alone would indicate a relatively young age, and a radiocarbon date on a piece of charcoal suggested an age of about 13,000 years. The skeleton, and particularly the skull and jaw, was studied in 1971 by Don Brothwell, my predecessor at the Natural History Museum, and he argued that while the specimen could be related to recent populations in West Africa, it actually looked rather different from them. I studied the skull for my Ph.D., with surprising results. I also found that it did not closely resemble recent African populations, but in its long and low shape it was actually closer to early moderns such as those from Skhul, and even to more primitive specimens such as Omo 2. This was decidedly odd for such a young skeleton, and so I recently collaborated in a new study of the specimen with the archaeologist Philip Allsworth-Jones, the dating expert Rainer Grün, and the anthropologist Katerina Harvati. We first checked with Thurstan Shaw whether there were any hints that the skull could have been much older than previously suggested, and there were none. With the help of the Nigerian archaeologist Philip Oyelaran, I obtained a fragment of bone from the skeleton and passed it to Grün in order to check its age directly. His determination from a direct uranium-series age estimate is that the bone is unlikely to be older than 20,000 years, consistent with the stratigraphy and associated archaeology and radiocarbon date.

Finally, could Brothwell and I have been wrong about the unusual shape of the skull? Harvati used state-of-the-art geometric morphometric scanning techniques on an exact replica of the skull (which is now in Nigeria) and found, as we did, that it was quite distinct from recent African crania, and indeed from any modern specimen in her comparative sample. Her results placed the skull closest to late archaic African fossils such as Ngaloba, Jebel Irhoud, and Omo 2—all thought to be at least 140,000 years old. So what does this mean? Because of the poor preservation of Pleistocene bones in West Africa, we have no other data on the physical form of the inhabitants of the region during the whole of the Pleistocene, so we have to be careful in interpreting an isolated specimen such as Iwo Eleru. But it does not seem to be diseased or distorted, and does indeed seem to indicate that Africa contained archaic-looking people in some areas when, and even long after, the first modern-looking humans had appeared. Support for this view comes from the work of the anthropologist Isabelle Crevecoeur. Her restudy of the numerous Ishango fossils from the Congo showed that these Later Stone Age humans were similar to Iwo Eleru not only in age but also in the surprisingly archaic features found in their skulls, jaws, and skeletons.

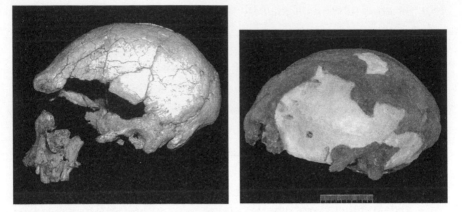

African fossils Ngaloba (Laetoli H.18, *left*) and Iwo Eleru (*right*).
They resemble each other, despite Iwo Eleru being less than 20,000 years old,
compared to Ngaloba's 140,000-year-old archaic features.

Africa today has the greatest internal genetic variation of any inhabited continent, and its skull shapes show the highest variation. This is
usually attributed to its greater size, larger ancient populations, and deepest time lines for humanity. But could those time lines go back even farther
than we thought? Did the early modern morphology evolve gradually and
then spread outward from a region like East Africa, completely replacing
archaic forms within Africa and then outside (as mtDNA data would suggest)? Or could there have been a version of assimilation or multiregional
evolution within Africa, with modern genes, morphology, and behavior
coalescing from partly isolated populations across the continent? Given
its huge size, complex climates, and patchworks of environments, Africa
could have secreted distinct human populations just as easily as the rest
of the inhabited world. So was the origin of modern humans there characterized by long periods of fission and fusion between populations, rather
than representing a sudden single event? And was the replacement of
the preceding late archaic peoples not absolute, so that they were partly
absorbed by the evolving moderns rather than completely dying out? In
which case, did early *Homo sapiens* forms, and even the preceding species
Homo heidelbergensis, survive alongside descendant modern humans?

This might explain archaic aspects in the shape of the Herto, Omo 2,
and Iwo Eleru crania. In part they resemble archaic crania like Broken

Hill, assigned to *Homo heidelbergensis,* so is this mosaic anatomy just a primitive retention from more ancient ancestors, or is it a sign of gene flow from contemporaneous African populations that still retained such features? My gut feeling is that some (but clearly not all) of the "ancient" DNA markers being picked up outside of Africa and used to argue for gene flow from non-African archaics will turn out to be traces of admixture that had actually happened in Africa. (A good example of this is the microcephalin gene discussed earlier.) Those traces were then carried from there in the modern human dispersal(s), followed by the operation of selection and drift on those populations, producing frequency changes in the genes when comparing the groups with each other and with their African counterparts. So while some archaic genes certainly were picked up by interbreeding outside of Africa, some were also acquired before the exodus, and yet others could even have been added in Africa, after it.

We still lack the amount of genetic data for African populations that we have for people from Europe and North America, but Africa is beginning to catch up. Charla Lambert and Sarah Tishkoff analyzed thousands of samples to reveal several deep and ancient population clusters, and, as we saw, Michael Hammer and his colleagues found evidence of archaic genes in three samples of moderns, but especially in West Africans. Now they have taken this work further by analyzing about half a million bits of genetic coding in samples of Mandenka (Senegal), Biaka pygmies (Central African Republic), and San (Namibia). They found strong evidence for split times of more than 100,000 years, predating the exodus from Africa, and they detected evidence of ancient admixture (with unknown "archaic" human groups) in both the Biaka and San. Philipp Gunz and his colleagues also recognized this from the great variability they found in late archaic/early modern African crania, as this edited extract of their conclusions shows.

> Our fossil AMH [anatomically modern human] data suggest that before there was isolation by distance [=drift] from Africa, there already existed (at least temporally) isolation by distance within Africa. Seemingly ancient contributions to the modern human gene pool have been explained by admixture with archaic forms of *Homo*, e.g. Neanderthals. Although we cannot rule out such admixture, the proposed ancestral

population structure of early AMH suggests another underestimated possibility: the genetic exchange between subdivided populations of early AMH as a potential source for "ancient" contributions to the modern human gene pool. Any model consistent with our data requires a more dynamic scenario and a more complex population structure than the one implied by the classic Out of Africa model. Our findings on neurocranial shape diversity are consistent with the assumption that intra-African population expansions produced temporarily subdivided and isolated groups. Separated demes (population subdivisions) might have partly merged again, whereas others left Africa at different times and maybe using different routes, and still others probably also remigrated to Africa.

However, there is still one significant loose end to be tied up, and that is the size of the human group that left Africa and founded the populations of the rest of the world. The lower genetic variation of peoples outside of Africa indicates that this population was only a small subsample of its parent group, and some calculations from mtDNA have suggested it could have been as low as a few hundred individuals. Yet it could not have been too sparse, or it would not have been able to contain the variation that, as I just argued, included archaic genes picked up in Africa.

There are also some clues that the Out of Africa dispersal was more complicated than a single exodus. The possibility that there were two exits separated by a significant period of time is not consistent with the tight ancestral patterns of both mitochondrial and Y-DNA, unless the earlier dispersal (perhaps represented by samples like those from Skhul and Qafzeh) is now only represented in autosomal DNA, with the more ancient DNA from their mitochondria and Y-chromosomes being completely replaced during the later dispersal. But the possibility of two dispersals closer in time to each other seems to be supported by some intriguing interpretations of data from the X-chromosome, where the single X in males is inherited from the mother, whereas the two Xs in females are inherited one from each parent. Because of this sex-based asymmetry in the inheritance of X-chromosomes, it is possible to garner data on the relative sizes of ancestral populations of males and females. Given typical one-to-one pair bonding, the expectation would be that there were equal numbers of male and female forebears, whereas a pattern of polygamous mating could produce a surfeit of female ancestors. Some genetic data do suggest that a polygamous mating pattern of males either having harems or engaging in serial polygamy may have characterized our evolution in the longer term, but at the time of the Out of Africa exit something quite different may have happened.

If there are equal numbers of males and females in a population, there will be three X-chromosomes for every four autosomal chromosomes, since, as we saw, males carry only one X. Genetic drift (the chance process of change in gene proportions that operates most markedly in small populations) should thus occur more strongly on the X-chromosome than on the other chromosomes, in the expected ratio of 4:3. Using a database of over 130,000 SNPs in the chromosomes of people from West Africa, Europe, and East Asia, the geneticists Alon Keinan, David Reich, and their colleagues calculated the actual amount of genetic drift that had occurred in the different populations. The West African sample (which should better represent the population structure within Africa before the exodus) met the expectations of approximately equal numbers of male and female ancestors. But the ratios in the non-African samples suggested accentuated drift in their X-chromosome DNA, indicating a smaller number of mothers than fathers around the time of the Out of Africa dispersal. Thus males must have dominated the population structure during or immediately after the exodus from Africa, probably across a time span of thousands of years.

If Keinan and Reich were correct, something decidedly odd must have happened about 55,000 years ago in northeast Africa or the adjoining Levant or Arabia—but what? There are several possible explanations, but their favored one is that at least two sequential founding populations were involved in the African exodus. The first provided the female representation, and thus the ancestral mtDNA of non-Africans (haplogroup L3 or its first M and N descendants). But the males in that first wave were somehow outreproduced or even replaced by new males, in one or several succeeding waves of dispersal, thus increasing the total male representation compared with females, and by inference involving a replacement of the original Y-chromosomes. One possibility is that subsequent bands of men violently replaced the original males and then mated with their females; if so, I wonder whether this might mark the arrival of projectile points in the region (see chapter 6). Alternatively could the process have been more gradual, with the new men holding a reproductive advantage over the original males—perhaps the development of a high-status elite or the export of a powerful new religion with male shamans, who gained privileged sexual access to the women?

These are just speculations, of course, and Keinan and Reich's scenario needs confirmation from further genetic analyses, but it will be interesting to see if new data emerge to support or disprove these conjectures. The possibility of multiple small bands of modern humans leaving Africa does increase the likelihood that there were enough people to carry some of those archaic African genes within them, in which case we do not always have to invoke the impact of hybridization with species outside of Africa to explain their presence. But equally, there clearly were at least occasional episodes of hybridization outside of Africa, with (in decreasing order of probability and increasing evolutionary distance) Neanderthals and other descendants of *heidelbergensis* (Denisovans?), surviving *erectus* in southeast Asia, and perhaps even (given the vagaries of human behavior) Hobbits in Flores!

So where does this added complexity and evidence of interbreeding with Neanderthals and Denisovans leave my favored Recent African Origin model (RAO)? Has it been disproved, in favor of the Assimilation or Multiregional models, as some have claimed? I don't think so, and to give a better perspective on this, I think we should revisit some of my first discussions about the early days of RAO, using a diagram (see p. 266). This compares different models of recent human evolution in terms of the extent of an African vs. non-African genetic contribution to present-day humans worldwide. On the left we have a pure Recent African Origin, with the total replacement of non-African genes, and at the other extreme we can envisage models that give Africa no place in the evolution of modern humans. (In chapters 1 and 3 I mention Clark Howell's proposal for a Middle East center of origin, and Christy Turner's for southeast Asia.) In between the extremes we can position "mostly Out of Africa" models like Günter Bräuer's toward the left-hand side and classic Multiregionalism (which gives Africa no special role—see below) toward the right. Somewhere near the center is Fred Smith and Erik Trinkaus's Assimilation model. If archaic gene flow was the rule rather than the exception, there could have been as much as a 50 percent non-African genetic input; however, depending on the extent of archaic gene flow envisaged, the Assimilation model could approach either Günter's Out of Africa + Hybridization model or classic Multiregionalism.

Back in 1970, no scientists held the view that Africa was the evolutionary home of modern humans; the region was considered backward and largely irrelevant, with the pendulum of scientific opinion strongly swinging toward non-African, Neanderthal Phase, or Multiregional models. Twenty years later, the pendulum was starting to move in favor of RAO, as fossil evidence began to be increasingly reinforced by the clear signals of mitochondrial and Y-DNA. The pendulum swung even farther toward a pure RAO with growing fossil, archaeological, and genetic data, including the distinctiveness of the first Neanderthal DNA sequences recovered in the late 1990s.

Now, the advent of huge amounts of autosomal DNA data, including the Neanderthal and Denisova genomes, has halted and even reversed that pendulum swing, away from an absolute RAO, and I would say we are looking at an RAO model that most resembles Bräuer's early formulation (Out of Africa + Hybridization) or a version of the Assimilation model of Smith and Trinkaus. If the evidence for archaic assimilation remains modest and restricted to Africa and the dispersal phase of modern humans from Africa, constituting less than 10 percent of our genome, I think "mostly Out of Africa" is the appropriate designation—and, for me, that is still RAO. I would have been delighted with that level of support for an African origin for *Homo sapiens* during the fierce arguments of the 1990s, when opponents still claimed Africa had no special role, so I'm more than happy with it now. And of course RAO is not just about the African origins of our shared modern morphology and most of our genes; it is also about the origins of our shared patterns of modern behavior.

Given that interbreeding seemingly did happen between modern and archaic humans, both in and out of Africa, does this mean that we should now abandon the different species names and lump all the fossils of the last million years or more as *Homo sapiens*, as some suggest? I think that if the hybridization events prove to have been widespread in time and space, we may well have to do that, but I don't think we are at that point

Which model is "right"?

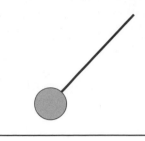

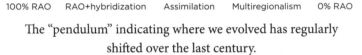

100% RAO   RAO+hybridization   Assimilation   Multiregionalism   0% RAO

The "pendulum" indicating where we evolved has regularly
shifted over the last century.

yet. There are still good scientific reasons to give populations that had long
and (relatively) separate evolutionary histories different names—species
or otherwise. As we saw, we can measure the amount of morphological
variation in primate species today, and then compare it with the differences between, say, *Homo erectus* and *heidelbergensis* skulls, or between
those of Neanderthals and moderns. This shows they are distinct enough
to be classed as different species, whether or not they meet the biological
species criterion of no interbreeding (a standard that numerous recognized primate species today do not achieve).

But if we nevertheless proceed to merge even Neanderthals and modern humans, we end up with a *Homo sapiens* characterized by, for example, a high and rounded skull, and a long and low skull; by no continuous
brow ridge, and a strong continuous brow ridge; by a well-developed chin
even in infants, and no chin; by no suprainiac fossa in adults, and a suprainiac fossa in adults; by an inner ear of modern shape, and an inner ear
of Neanderthal shape; by a narrow pelvis with a short thick superior pubic
ramus, and a wide pelvis with a long thin superior pubic ramus—and so
on. The disparate nature of *Homo sapiens* would become even more extreme
if we started to add in the features of species like *heidelbergensis*, *antecessor*, and *erectus*.

The merging of *sapiens* and *erectus* has been regularly proposed by
champions of Multiregionalism, and if our modern genes do come from

more than one region, why hasn't the Multiregional theory been proved correct, as its remaining supporters are now asserting? To deal with that claim, I think it is worth reminding ourselves what classic Multiregionalism actually proposed. Here is a quotation from a paper written in 1994 by Milford Wolpoff and four other prominent advocates of the model at that time.

> The evolutionary patterns of three different regions show that the earliest "modern" humans are not Africans and do not have the complex of features that characterize the Africans of that time or any other . . . There is no evidence of specific admixture with Africans at any time, let alone replacement by them . . . There is indisputable evidence for the continuity of distinct unique combinations of skeletal features in different regions, connecting the earliest human populations with recent and living peoples.

This model gave Africa no special place in our evolution and claimed specific connections in individual features between *Homo erectus* fossils more than a million years old in each region and humans in the same regions today. Hopefully the evidence I have presented in this book shows that these particular views have been pretty comprehensively shown to be false.

But still, if we do have significant components of archaic genes in our makeup, why have these not featured strongly in genetic analyses until now? I mentioned there were hints of them from time to time, but the reality is that if our archaic component represents only about 5 percent of our genome, then ninety-five times out of a hundred this would not show up in individual genetic marker studies, just as it did not register in our mtDNA and Y-chromosomes. It has taken much larger genome sweeps, and comparisons with actual archaic genomes, to show up the signs of ancient hybridization.

The big picture is that we are predominantly of recent African origin, so is there a special reason for this? Overall, I think that the preeminence of Africa in the story of modern human origins was a question of its larger geographic and human population size, which gave greater opportunities for morphological and behavioral variations, and for innovations to develop and be conserved, rather than the result of a special evolutionary pathway. "Modernity" was not a package that had a unique African origin in one time, place, and population, but was a composite whose elements appeared at different times and places and were then gradually assembled to assume the form we recognize today. But if that was so, is the assembly over, and has the evolution of modern humans finally ended? Are we the finished product and in control of our destiny, or are many of the same processes that operated in our past continuing to affect us now and in the future? The scientist and author Stephen Jay Gould's view was clear: "There's been no biological change in humans in 40,000 or 50,000 years. Everything we call culture and civilization we've built with the same body and brain."

When I give public lectures, I am invariably asked where evolution will lead us, and what humans will look like in the future, and equally invariably I try to avoid answering such tricky questions. I have, though, taken a different view in public from Gould and my geneticist friend Steve Jones over whether human evolution is over. Jones suggests that modern culture and its benefits like medical care removed the power of natural selection to affect humans, since virtually everyone now reaches reproductive age. I disagree because, first, changes in our genome are occurring all the time, whether we can detect them or not; some calculations suggest that each of us could have about fifty new mutations compared with our parents' DNA. Second, life in the developed world has its own differential costs in terms of reproduction and health, with the general availability of contraception, but also of junk food, alcohol, and drugs. Third, and even more significant, at least a quarter of the world's population is still denied the benefits of decent health care and the necessities of healthy living conditions and diets. Thus selection is operating strongly on those billions of people, and I cannot see that stopping any time soon. From my perspective, evolution is certainly still working away on *Homo sapiens*, and there is even evidence that its effects have accelerated rather than diminished over the last 10,000 years, as we will see below.

Science fiction images of humans of the future often show us with huge brains but, as we saw, big brains are not necessarily the best brains—witness the extinct Neanderthals—and if anything our brains have actually shrunk in size over the last 20,000 years. In practical terms, unless the process of birth is bypassed, our brain size is already at the limit at which the female pelvis can cope with delivery. Then there is the sheer cost in energy of running a big brain, and the evidence that larger brains are not necessarily as efficient at some tasks. And anyway, so much of our memorizing and thinking is done externally now—in the brains of other people or in the processors of our computers. All of these factors could be responsible for our shrinking brains, as well as more mundane factors like an overall reduction of body size compared with our Paleolithic ancestors.

More realistically for our future evolution, there is the prospect of genetic engineering, which is already happening on a small scale. Genetic counseling is available to advise potential parents about harmful DNA mutations that could be passed on to their children, and to give them the choice about whether to proceed. As this becomes more common and wider in its reach, future gene pools will be affected. Even more ambitiously, gene therapy could be applied to a faulty organ in the body, and germ-line therapy could plant a permanent change in the genome of an unborn baby. There are formidable ethical questions to be addressed here, not to mention the scientific ones. For example, we know that the actions of genes are often interrelated, and that a single gene may perform more than one function. So great care would be needed to ensure that the targeted change in DNA achieved only what was intended. And the social consequences of even giving people the simple choice of a male or a female child are enormous, let alone providing opportunities to enhance that child's beauty, talents, or intelligence.

Most of that is still science fiction, and perhaps it is for the best that some elements will always remain so. But for the last 10,000 years, selection seems to have been engineering people to cope with massive lifestyle changes. When humans expanded into novel environments over the last 50,000 years, including into rain forests in Africa and new habitats in Eurasia, Australasia, and the Americas, they encountered fresh challenges and had to adapt both physically and culturally. The physical adaptations ranged from gross changes in body size or shape, down to immunological responses to a host of new pathogens. And over the last

20,000 years these have also included distinct mutations in Europe and Asia for depigmentation, to assuage the lower levels of sunlight, as well as the spread of blue eyes in western Eurasia—although this latter change might equally have come from cultural selection.

Culture, rather than slowing changes in our DNA, may well have provided the means to speed them up. This is the view of a growing number of geneticists and anthropologists, including Henry Harpending, Gregory Cochran, John Hawks, Anna Di Rienzo, Pardis Sabeti, Sharon Grossman, Ilya Shylakhter, and Kevin Laland. They argue that profound changes in human lifestyles over the last 10,000 years—with the moves to pastoralism, agriculture, and urbanization—would have had equally profound evolutionary effects. With the consequent huge increase in human numbers, there are clear parallels with the relationship between demography and innovation: a larger population will not only have more mutations, and more beneficial mutations, but also provide better chances for them to be conserved and disseminated. And the fact that farming also entailed self-induced changes in societies, diets, and environments (not all of them beneficial to everyone) would have ensured that selection remained a powerful force for evolutionary change.

Ten thousand years ago, as agriculture was taking off in its west and east Asian cradles, the hunter-gatherer populations of the world probably numbered only a few million people, and in many areas they must have been thinly scattered. The estimated figure only 8,000 years later was over 200 million, and following the industrial revolution and the advent of measures like vaccination, our numbers are now soaring toward 10 billion. The huge increase between 10,000 and 2,000 years ago would have ensured a proportionate increase in mutations, including potentially favorable ones, and, provided population density was high (which it was in many agricultural and subsequent urban communities), any genetic changes had the potential to spread rapidly. As people acquired stable food supplies through farming, they settled down in increasingly large communities, but with this change there were also many downsides. Unsanitary living conditions and densely packed communities were ripe for parasites and epidemic diseases like smallpox, cholera, and yellow fever, while the clearance of forests and the use of irrigation led to the spread of malaria across much of the tropics and subtropics. Overreliance on one or two staple foods

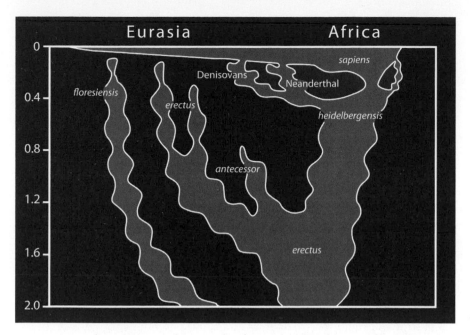

An evolutionary tree showing the geographic distribution of humans
and humanlike relatives from the last 2 million years. Note the complexity
of relationships now implied by the latest genetic data.

also meant that the benefits of a broader hunter-gatherer diet were lost, and,
for many, hard labor in the fields wore out bodies prematurely. Societies
and technologies had to keep up with the changes too. People were thrown
together and socializing in larger numbers, with the growth of task specializations, and disparities in wealth, status, and, no doubt, reproductive
success.

All of these upheavals should have provided fertile ground for evolution to operate, and so many groups of geneticists have been combing the
human genome for signs of this. The methods are known as *genome-wide*
or *whole-genome association studies*, where a correlation is sought between
genes and particular traits, whether these are physical, such as skin color
or height, or physiological, such as susceptibility to a disease. Of course
such studies must take into account environmental influences, as well as
the complexity of gene expression, since a particular end result may come
from the interaction of several different genes, rather than just one. A major
source for association studies has been the International Haplotype Map,

which has provided data on millions of SNPs in 270 people of European, Nigerian, Chinese, and Japanese descent. These single letter mutations are inherited within larger sequences of DNA, and segments break down over time as a result of the remixing of the DNA on our chromosomes with every new generation. New mutations can be spotted, and their age can be estimated by the amount of mixing that has occurred around them.

Sure enough, the signals of recent selection not only were there but were also very strong, acting on perhaps 20 percent of our genes. Some could be directly related to the changes induced by farming, linked to new diets, such as the gene for lactase. This is an enzyme that allows infants to digest lactose (milk sugar) when nursing, but it usually switches off during childhood, so that many adults are lactose-intolerant. However, in the last 10,000 years, separate genetic changes occurred in East Africa and regions of western Eurasia that prevented the lactase gene switching off, meaning that adults (about 80 percent in the case of Europeans) can comfortably digest milk from livestock. Populations elsewhere who lack the mutations, such as East Asians and Native Australians and Americans, are still only able to drink milk comfortably as babies. Meanwhile, mutations have evolved to allow the digestion of other "new" carbohydrates in the diet in West Africa (for the sugar mannose) and East Asia (mannose and sucrose). And there have also been changes in a gene that codes for salivary amylase (which helps to digest starch), both in its structure and in the number of copies of the gene in many individuals. Examples of recent selection in human genes have been known for many years in connection with protection from malaria, and at least twenty-five different examples have now been detected. Because the malarial parasite is transmitted in the bloodstream, many human defenses originate in the blood, such as mutations in the hemoglobin gene, which carries oxygen, or in the enzyme G6PD. And blood groups have responded too, with an entirely new one—Duffy—seemingly selected specifically to combat the disease. Many further changes seem to be related to resistance to infectious diseases such as tuberculosis, and 10 percent of Europeans have been fortuitous in carrying mutations that have apparently been selected to resist smallpox; they also seem to confer resistance to HIV.

Other recent changes may be related to the changing social conditions brought by agricultural life. In chapter 6, we mentioned mutations in the gene for the cholesterol-transporting apolipoprotein E that seem to lower the risk of many age-related conditions such as coronary disease, and there are at least fourteen other recently mutated genes that are linked with conditions most expressed in the old, such as cancers and Alzheimer's. Considering the crucial importance of extended families for both hunter-gatherers and farmers, selection seems to have been working on the survival of people past reproductive age as well, given the consequent social benefits. But a possible downside for social harmony from higher population densities is the greater potential for adultery, and this may be reflected in widespread but regionally distinct mutations controlling the quantity and vigor of human sperm—perhaps indicative of "sperm competition," caused when a woman partners more than one man within a day or so. Perhaps some of the one hundred or so recent mutations in brain neurotransmitters concerned with mood and demeanor have correspondingly been selected to cope with the social consequences of our large population numbers and the possible resultant tensions.

Those neurotransmitters are only a part of our changing genome as far as the brain and senses are concerned. Although this is a highly controversial area, it is likely that selection has favored different behaviors and cognitive abilities as modern humans have diversified in different environments and social complexities. With the development of specialized occupations and their associated skills, selection may have increasingly come into play. For example, the need to work out stocks of cereals or animals, followed by the rise of trading and the arrival of money, would all have encouraged selection for mathematical abilities. And the increasing complexity of communication in small or ever-larger groups may be marked by recent mutations in genes that produce proteins for the cilia of our inner ears and the membrane that coats them, as well as one that helps to build the actual bones of the middle ear, which transmits sonic frequencies. The fact that different mutations are found when comparing Chinese and Japanese, Europeans and Africans suggests that selection might even have been tracking the evolution of different languages and their most characteristic sounds. Sight, too, may have been under recent selection in East Asia—mutations in the protocadherin-15 gene there affect the workings of both inner ear cells and photoreceptors in the retina.

But to return to the question posed earlier, it appears that human evolution, at least in terms of changes in individual DNA sequences, has accelerated rather than slowed or stopped over the last 10,000 years. Indeed, some calculations suggest that it is now happening a hundred times faster than it did since we split from the lineage of chimpanzees, probably more than 6 million years ago. About 7 percent of human genes seem to have mutated recently in some populations, the majority within the last 40,000 years, and particularly within the last 10,000 years. Some caution has to be injected here, since geneticists like Sarah Tishkoff and Mark Stoneking have pointed out that the expansion of human populations might have increased rare variants by chance alone, so the functional benefit of the genetic change needs to be properly demonstrated—as it can be in many cases. Additionally, and perhaps more seriously, the constant loss and overwriting of changes in our DNA mean that some ancient signals of genetic change—during the Middle Stone Age, for example—have been lost or are difficult to detect now. Thus we have a biased signal for the last 10,000 years or so, because this is the very period when we have the most chance of recognizing novel mutations.

Fortunately this is a fast-moving area of science, and a lot of new data will be arriving to resolve this question in the next few years—including a thousand complete human genomes from around the world. Pardis Sabeti, as well as singing in a rather good rock band, has worked with her colleagues on a new method that combines three tests for multiple signals of selection, and which has the potential to increase the resolution of scans for recently selected DNA as much as a hundredfold. She is also researching something important that we haven't touched on—not every genetic change involves our DNA. Ribonucleic acid (RNA), like DNA, consists of long chains of nucleotides, but these chains are usually single-stranded in our cells. Different types of RNA are central to protein synthesis and to the regulation of gene expression, and thus RNA—which also mutates—forms another subject and agent of evolutionary change. This is part of a growing body of data concerning inheritance that lies beyond the genetic code of DNA, constituting the field of epigenetics (from the Greek, meaning "over or above genetics"). This is a fast-developing area of research that will not replace the current focus on DNA, but certainly provides additional ways of looking at inheritance and evolution. Here, short-term environmental changes may have an impact on bodily form and function beyond changes purely in our DNA—for example, via histone proteins that make up part of the chromosomes, or through the modifications that viruses or prions may inflict on us.

Finally, although this discussion of recent changes in human DNA has constantly referred to selection, we should bear in mind that selective changes may not benefit everyone; there can be winners and losers, as there has been with the rise of sickle-shaped cells in the blood of African-derived populations. Sickling has benefited those who are heterozygous for the sickle-cell gene (that is, they have only one copy of it) by conferring some immunity against the malarial parasite. But without medical intervention, those born with two copies of the gene will be highly anemic and will die prematurely. The frequency of a mutation in the leptin receptor gene has increased dramatically in East Asia, linked with changes in the body mass index and a tendency to store fat. This may have been beneficial for adaptation to colder climates but now is a cause of high blood pressure and obesity. Some researchers have also argued that long and stressful sea voyages, whether forced in the case of the slave trade or voluntary in the case of the colonization of Polynesian islands, would have selected physiques and physiologies that were best able to survive the rigors of those journeys. The survivors then went on to found much larger populations who now live under very different conditions, perhaps explaining the prevalence of salt-sensitive hypertension in American blacks, and of diabetes and obesity in parts of Oceania. Similarly, as the anthropologist Peter Ellison pointed out, it is possible that the apparent increasing frequency of conditions like autism, schizophrenia, allergies, asthma, autoimmune diseases, and reproductive cancers are the modern downside of genetic changes that were beneficial under more ancient human environments and lifestyles. These comparisons between the past and the present are the basis of a whole new field of science called *evolutionary medicine*.

It is not always clear what the precise agent of selection has been in the past, beyond differential reproductive success. Where disease is concerned, it is obvious that this will cause direct natural selection through the reduced fertility or death of those whose natural (inherited) defenses are unable to cope with the condition in question. But exposure to the pathogens may be reduced or increased by particular human behaviors (think of the use of condoms, which act as contraceptives, but also combat the spread of HIV). Thus many of these changes probably lie in the realm of complex interactions between the natural environment and ones we have created through our diversity of human cultures. And this brings us back to one of Darwin's favored evolutionary mechanisms, as highlighted by the full title of his second most famous book: *The Descent of Man, and Selection in Relation to Sex*. It is evident that, as Darwin proposed, some of these changes could be ascribed to human sexual/cultural selection, where habitual preferences in mating could steer evolution in a particular direction. This might well include some of the regional ("racial") differences in appearance, as Darwin suspected, and equally some of the changes in brains and behavior. Stature is a case in point; it is a complex trait, but one with a high heritability. There is evidence that stature (as long as it does not become excessive) is linked to both fecundity and wealth in the developed world, and studies of the selection of sperm donors suggest that women prefer taller donors, which will in turn lead to taller offspring.

All of this would have fascinated Darwin. When he was alive, the hard data about our origins could have been packed up in a small suitcase, and while in many ways he started the writing of the book of human evolution, all he could achieve was the equivalent of drafting some chapter titles and words and sentences scattered through it. Since then we have learned so much about our early history, and many more words, sentences, and paragraphs of our story are now in place. Some chapters are fairly complete, such as the ones about building complete human and chimpanzee genomes, with the Neanderthal and Denisovan chapters following now. And yet the writing of other chapters has hardly begun, such as the ones about how our brains really work, who were the first peoples of the Indian subcontinent, the early history of the Hobbit in southeast Asia, and who was living in West Africa for most of prehistory.

Certainly, until we have a dated fossil, archaeological, and environmental record from many more regions to match the quality of the ones we have from western Europe, and are beginning to have from places like eastern and southern Africa, we cannot even guess how the book of our evolutionary history will look when it nears completion. Paleoanthropology is such a fast-moving and fast-developing science that even some of what is already written in that book will need to be corrected, or perhaps even deleted altogether, including my own contributions, no doubt. The process of writing this book has led me to a greater recognition of the forces of demography, drift, and cultural selection in recent human evolution than I had considered before. And while I have been writing it, new genetic data have emerged to show that we *Homo sapiens* are not purely derived from a recent African origin. But this dynamism is what makes studying human evolution so fascinating, and science is not about being right or wrong, but about gradually approaching truth about the natural world.

When Darwin died and was granted the honor of being buried in Westminster Abbey, there were many rich tributes to the man and his work, as this example shows.

> Mr. Darwin has left as broad and deep a mark upon Psychology as he has upon Geology, Botany, and Zoology. Groups of facts which previously seemed to be separate, are now seen to be bound together in the most intimate manner; and some of what must be regarded as the first principles of the science, hitherto unsuspected, have been brought to light. If the proper study of mankind is man, Mr. Darwin has done more than any other human being to further the most desirable kind of learning, for it is through him that humanity in our generation has first been able to begin its response to the precept of antiquity—know thyself.

That last phrase harks back to ancient Greece but was also Linnaeus's directive in describing the species he named *Homo sapiens*. Knowing thyself, for me, has meant a journey from measuring fossil skulls in European museums forty years ago to looking at almost every aspect of our origins. Knowing ourselves has meant recognition that becoming "modern" is the path we perceive when looking back on our own evolutionary history. That history seems special to us, of course, because we owe our very existence to it. Those figures of human species (usually males) marching boldly across the page have illustrated our evolution in many popular articles, but they have wrongly enshrined the view that evolution was simply a progression leading to us, its pinnacle and final achievement. Nothing could be further from the truth. There were plenty of other paths that could have been taken; many would have led to no humans at all, others to extinction, and yet others to a different version of "modernity." We can only inhabit one version of being human—the only version that survives today—but what is fascinating is that paleoanthropology shows us those other paths to becoming human, their successes and their eventual demise, whether through failure or just sheer bad luck. Sometimes the difference between failure and success in evolution is a narrow one, and we are certainly on a knife edge now as we confront an overpopulated planet and the prospect of global climate change on a scale that humans have never faced before. Let's hope our species is up to the challenge.

Sources and Suggested Reading

General Reading
Boyd, R., and J. Silk. *How Humans Evolved.* 5th ed. Norton, New York, 2009.

Cartmill, M., and F. Smith. *The Human Lineage.* 2nd ed. Wiley-Blackwell, Hoboken, NJ, 2009.

Darwin, C. The Complete Work of Charles Darwin Online. http://darwin-online.org.uk.

Fagan, B. *Cro-Magnon: How the Ice Age Gave Birth to the First Modern Humans.* Bloomsbury Press, London, 2010.

Johanson, D., and K. Wong. *Lucy's Legacy: The Quest for Human Origins.* Harmony Books, New York, 2009.

Klein, R. G. *The Human Career.* University of Chicago Press, Chicago, 2009.

Lewin, R., and R. A. Foley. *Principles of Human Evolution.* Blackwells, Oxford, 2003.

Lockwood, C. *The Human Story.* Natural History Museum, London, 2007.

Potts, R., and C. Sloan. *What Does It Mean to Be Human?* National Geographic, Washington, DC, 2010.

Stringer, C. *Homo britannicus.* Allen Lane, London, 2006.

———. "Modern Human Origins—Progress and Prospects." *Philosophical Transactions of the Royal Society, London (B)* 357 (2002), 563–79.

Stringer, C., and P. Andrews. *The Complete World of Human Evolution.* Thames & Hudson, London, 2005.

Stringer, C., and C. Gamble. *In Search of the Neanderthals.* Thames & Hudson, London, 1993.

Wood, B. *Human Evolution (a Brief Insight).* Sterling, New York, 2011.

Zimmer, C. *Smithsonian Intimate Guide to Human Origins.* Harper, New York, 2007.

1. The Big Questions
Beaumont, P. B., H. De Villiers and J. C. Vogel. "Modern man in sub-Saharan Africa prior to 49,000 years B.P.: A review and evaluation with particular reference to Border Cave." *South African Journal of Science* 74 (1978), 409–19.

Brace, C. L. "The fate of the 'Classic' Neanderthals: A consideration of hominid catastrophism." *Current Anthropology* 59 (1964), 3–43.

Bramble, D. M., and D. E. Lieberman. "Endurance running and the evolution of *Homo*." *Nature* 432 (2004), 345–52.

Bräuer, G. "The 'Afro-European *sapiens* hypothesis' and hominid evolution in east Asia during the Middle and Upper Pleistocene." *Courier Forschungsinstitut Senckenberg* 69 (1984b), 145–65.

Campbell, B. G. "Conceptual progress in physical anthropology: Fossil Man." *Annual Review of Anthropology* 1 (1972), 27–54.

Cann, R. L., M. Stoneking and A. C. Wilson. "Mitochondrial DNA and human evolution." *Nature* 329 (1987), 111–12.

Cavalli-Sforza, L. L., A. Piazza, P. Menozzi and J. Mountain. "Reconstruction of human evolution: Bringing together genetic, archaeological, and linguistic data." *Proceedings of the National Academy of Sciences USA* 85 (1988), 6002–6.

Clark, D. "Africa in prehistory: Peripheral or paramount?" *Man* 10 (1975), 175–98.

Coon, C. S. *The Origin of Races.* Alfred A. Knopf, New York, 1962.

Dart, R. A. "*Australopithecus africanus*: The man-ape of South Africa." *Nature* 115 (1925), 195–99.

Darwin, C. *The Descent of Man, and Selection in Relation to Sex.* John Murray, London, 1871; Penguin Classics, London, 2004.

———. *On the Origin of the Species by Means of Natural Selection, or the Preservation of Favoured Races in the Struggle for Life.* John Murray, London, 1859.

Gröning, F., J. Liu, M. J. Fagan and P. O'Higgins. "Why do humans have chins? Testing the mechanical significance of modern human symphyseal morphology with finite element analysis." *American Journal of Physical Anthropology* (2010) (doi: 10.1002/ajpa.21447).

Howell, F. C. "Upper Pleistocene men of the southwest Asian Mousterian." In G. H. R. von Koenigswald, ed., *Neanderthal Centenary, 1856–1956*, pp. 185–98. Kemink en Zoon, Utrecht, 1958.

Howells, W. W. "Explaining modern man: Evolutionists versus migrationists." *Journal of Human Evolution* 5 (1976), 477–96.

Hrdlička, A. *The Skeletal Remains of Early Man.* Smithsonian Institution, Washington, DC, 1930.

Klein, R. G. *The Human Career.* University of Chicago Press, Chicago, 2009.

Leakey, R. E. F., K. W. Butzer and M. H. Day. "Early *Homo sapiens* remains from the Omo River region of Southwest Ethiopia." *Nature* 222 (1969), 1132–38.

Linnaeus, C. *Systema Naturae.* 10th ed. Vol. 1. Holmiae, Salvii, 1758.

McCown, T. D., and A. Keith. *The Stone Age of Mount Carmel.* Vol. 2. Clarendon, Oxford, 1939.

Notton, D., and C. Stringer. "Who Is the Type of *Homo sapiens*?" (2010). http://iczn.org/content/who-type-homo-sapiens.

Reader, J. *Missing Links: The Hunt for Earliest Man.* Collins, London, 1990.

Schwartz, J. H., and I. Tattersall. "Fossil Evidence for the Origin of *Homo sapiens*. *Yearbook of Physical Anthropology* 53 (2010), 94–121.

Shipman, P. *The Man Who Found the Missing Link: Eugène Dubois and his Lifelong Quest to Prove Darwin Right.* Simon & Schuster, New York, 2001.

Smith, A. *Systematics and the Fossil Record: Documenting Evolutionary Patterns.* Oxford, Blackwell, 1994.

Smith, F. H., I. Janković and I. Karavanić. "The assimilation model, modern human origins in Europe, and the extinction of Neandertals." *Quaternary International* 137 (2005), 7–19.

Stringer, C. "Out of Africa—a Personal History." In M. H. Nitecki and D. V. Nitecki, eds., *Origins of Anatomically Modern Humans*, pp. 151–72. Plenum Press, New York, 1994.

Stringer, C., and R. McKie. *African Exodus.* Cape, London, 1996.

Templeton, A. R. "The 'Eve' hypothesis: A genetic critique and reanalysis." *American Anthropologist* 95 (1993), 51–72.

Trinkaus, E. "Early Modern Humans." *Annual Review of Anthropology* 34 (2005), 207–30.

Weidenreich, F. "Facts and speculations concerning the origin of *Homo sapiens*." *American Anthropologist* 49 (1947), 187–203.

White, R. "Rethinking the Middle/Upper Paleolithic transition." *Current Anthropology* 23 (1982), 169–92.

Wolpoff, M., C. B. Stringer and P. Andrews. "Modern human origins." *Science* 241 (1988), 773–74.

Wolpoff, M. H., A. G. Thorne, F. H. Smith, D. W. Frayer and G. G. Pope. "Multiregional evolution: A world-wide source for modern human populations." In M. H. Nitecki and D. V. Nitecki, eds., *Origins of Anatomically Modern Humans*, pp. 175–99. Plenum Press, New York, 1994.

Wolpoff, M. H., W. X. Zhi and A. G. Thorne. "Modern *Homo sapiens* origins: A general theory of hominid evolution involving the fossil evidence from east Asia." In F. H. Smith and F. Spencer, eds., *The Origins of Modern Humans: A World Survey of the Fossil Evidence*, pp. 411–83. Alan R. Liss, New York, 1984.

2. Unlocking the Past

Aitken, M., C. B. Stringer and P. Mellars. *The Origin of Modern Humans and the Impact of Chronometric Dating.* Princeton University Press, Princeton, NJ, 1993.

Ambrose, S. H. "Did the super-eruption of Toba cause a human population bottleneck? Reply to Gathorne-Hardy and Harcourt-Smith." *Journal of Human Evolution* 45 (2003), 231–37.

Banks, W. E., F. d'Errico, A. Townsend Peterson, M. Masa Kageyama, A. Sima and M.-F. Sanchez-Goni. "Neanderthal extinction by competitive exclusion." *PLoS ONE* 3 (2008), 1–8.

Bradtmöller, M., A. Pastoors, B. Weninger and G.-C. Weninger. "The repeated replacement model—rapid climate change and population dynamics in Late Pleistocene Europe." *Quaternary International* (2010) (in press).

Bronk Ramsey, C., T. Higham, A. Bowles and R. Hedges. "Improvements to the pre-treatment of bone at Oxford." *Radiocarbon* 46 (2004), 155–63.

Castañeda, I. S., S. Mulitza, E. Schefuß, R. A. Lopes dos Santos, J. S. Sinninghe Damsté and S. Schouten. "Wet phases in the Sahara/Sahel region and human migration patterns in North Africa." *Proceedings of the National Academy of Sciences USA* 106 (2009), 20159–63.

d'Errico, F., C. T. Williams and C. Stringer. "AMS dating and microscopic analysis of the Sherborne bone." *Journal of Archaeological Sciences* 25 (1998), 777–87.

Drake, N. A., A. S. El-Hawat, P. Turner, S. J. Armitage, M. J. Salem, K. H. White and S. McLaren. "Palaeohydrology of the Fazzan Basin and surrounding regions: The last 7 million years." *Palaeogeography, Palaeoclimatology, Palaeoecology* 263 (2008), 131–45.

Finlayson, C., and J. S. Carrión. "Rapid ecological turnover and its impact on Neanderthal and other human populations." *Trends in Ecology and Evolution* 22 (2007), 213–22.

Gathorne-Hardy F. J., and W. E. H. Harcourt-Smith. "The super-eruption of Toba, did it cause a human bottleneck?" *Journal of Human Evolution* 45 (2003), 227–30.

Gowlett, J., and R. E. M. Hedges. *Archaeological Results from Accelerator Dating.* Oxford University School of Archaeology, Oxford, 1987.

Grün, R. "Direct dating of human fossils." *Yearbook of Physical Anthropology* 49 (2006), 2–48.

Grün, R., J. Brink, N. Spooner, L. Taylor, C. Stringer, R. Franciscus and A. Murray. "Direct dating of Florisbad hominid." *Nature* 382 (1996), 500–501.

Grün, R., and C. B. Stringer. "Electron spin resonance dating and the evolution of modern humans." *Archaeometry* 33 (1991), 153–99.

———. "Tabun revisited: Revised ESR chronology and new ESR and U-series analyses of dental material from Tabun C1." *Journal of Human Evolution* 39 (2000), 601–12.

Grün, R., C. Stringer, F. McDermott, R. Nathan, N. Porat, S. Robertson, L. Taylor, G. Mortimer, S. Eggins and M. McCulloch. "U-series and ESR analyses of bones and teeth relating to the human burials from Skhul." *Journal of Human Evolution* 49 (2005), 316–34.

Grün, R. C. B. Stringer and H. P. Schwarcz. "ESR dating of teeth from Garrod's Tabun Cave collection." *Journal of Human Evolution* 20 (1991), 231–48.

Jacobi, R. M., and T. F. G. Higham. "The early lateglacial re-colonization of Britain: New radiocarbon evidence from Gough's Cave, southwest England." *Quaternary Science Reviews* 28 (2009), 1895–913.

Jacobs, Z., and R. G. Roberts. "Advances in optically stimulated luminescence dating of individual grains of quartz from archaeological deposits." *Evolutionary Anthropology* 16 (2007), 210–23.

Jacobs, Z., R. G. Roberts, R. F. Galbraith, H. J. Deacon, R. Grün, A. Mackay, P. Mitchell, R. Vogelsang and L. Wadley. "Ages for the Middle Stone Age of Southern Africa: Implications for human behavior and dispersal." *Science* 322 (2008), 733–35.

Jones, S. C. "Palaeoenvironmental response to the ~74 Ka Toba ash-fall in the Jurreru and Middle Son valleys in southern and north-central India." *Quaternary Research* 73 (2010), 336–50.

Libby, W. F. *Radiocarbon Dating.* University of Chicago Press, Chicago, 1955.

McDougall, I., F. H. Brown and J. G. Fleagle. "Stratigraphic placement and age of modern humans from Ethiopia." *Nature* 433 (2005), 733–36.

Mellars, P., and J. French. "Tenfold population increase in Western Europe at the Neandertal-to-modern human transition." *Science* 333 (2011), 623–27.

Mercier, N., H. Valladas, O. Bar-Yosef, B. Vandermeersch, C. Stringer and J.-L. Joron. "Thermoluminescence date for the Mousterian burial site of Es-Skhul, Mt. Carmel." *Journal of Archaeological Science* 20 (1993), 169–74.

Millard, A. R. "A critique of the chronometric evidence for hominid fossils: I. Africa and the Near East 500–50 ka." *Journal of Human Evolution* 54 (2008), 848–74.

Müller, U. C., J. Pross, P. C. Tzedakis, C. Gamble, U. Kotthoff, G. Schmiedl, S. Wulf and K. Christanis. "The role of climate in the spread of modern humans into Europe." *Quaternary Science Reviews* 30 (2011), 273–79.

Oppenheimer, C. "Limited global change due to the largest known Quaternary eruption, Toba ca. 74 kyr BP?" *Quaternary Science Reviews* 21 (2002), 1593–1609.

Osborne, A. H., D. Vance, E. J. Rohling, N. Barton, M. Rogerson and N. Fello. "A Humid Corridor across the Sahara for the Migration 'Out of Africa' of early modern humans 120,000 years ago." *Proceedings of the National Academy of Sciences USA* 105 (2008), 16444–47.

Petraglia, M. D., R. Korisettar, N. Bolvin, C. Clarkson, P. Ditchfield, S. Jones, J. Koshy et al. "Middle Paleolithic assemblages from the Indian subcontinent before and after the Toba Super-Eruption." *Science* 317 (2007), 114–16.

Rampino, M. R., and S. H. Ambrose. "Volcanic winter in the Garden of Eden: The Toba supereruption and the late Pleistocene human population Crash." *Geological Society of America Special Paper* 345 (2000), 71–82.

RESET Project: http://c14.arch.ox.ac.uk/RESET/embed.php?File=index.html.

Robock, A., C. M. Ammann, L. Oman, D. Shindell, S. Lewis and G. Stenchikov. "Did the Toba volcanic eruption of ~74k BP produce widespread glaciation?" *Journal of Geophysical Research* 114 (2009), D10107 (doi: 10.1029/ 2008JD011652).

Stewart, J. R., M. van Kolfschoten, A. Markova and R. Musil. "The mammalian faunas of Europe during oxygen isotope stage three." In T. H. van Andel and S. W. Davies, eds., *Neanderthals and Modern Humans in the European Landscape During the Last Glaciation, 60,000 to 20,000 Years Ago: Archaeological Results of the Stage 3 Project*, pp. 103–29. McDonald Institute for Archaeological Research, Cambridge, 2003.

Stringer, C. B. "Dating the origin of modern humans." In C. Lewis and S. Knell, eds., *The Age of the Earth: From 4004 BC to AD 2002*, pp. 265–74. Geological Society, London, 2001.

———. "Direct dates for the fossil hominid record." In J. Gowlett and R. E. M. Hedges, eds., *Archaeological Results from Accelerator Dating*, pp. 45–50. Oxford University, Oxford, 1986.

Stringer, C. B., and R. Burleigh. "The Neanderthal problem and the prospects for direct dating of Neanderthal remains." *Bulletin of the British and Natural History Museum*, Geology Series 35 (1981), 225–41.

Stringer, C. B., R. Grün, H. Schwarcz and P. Goldberg. "ESR dates for the hominid burial site of Es Skhul in Israel." *Nature* 338 (1989), 756–58.

Stringer, C., R. Jacobi and T. Higham. "New research on the Kent's Cavern 4 maxilla, its context and dating." In C. Stringer and S. Bello, eds., *First Workshop of AHOB2: Ancient Human Occupation of Britain and Its European Context*, pp. 25–27. AHOB, London, 2007.

Stringer, C., H. Pälike, T. van Andel, B. Huntley, P. Valdes and J. Allen. "Climatic stress and the extinction of the Neanderthals." In T. H. van Andel and S. W. Davies, eds., *Neanderthals and Modern Humans in the European Landscape During the Last Glaciation*, pp. 233–40. McDonald Institute for Archaeological Research, Cambridge, 2003.

Timmreck, C., H.-F. Graf. S. J. Lorenz, U. Niemeier et al. "Aerosol size confines climate response to volcanic super-eruptions." *Geophysical Research Letters* 37 (2010), L24705 (doi: 10.1029/2010GL045464).

Williams, M. A. J., S. H. Ambrose, S. van der Kaars, C. Ruehlemann, U. Chattopadhyaya, J. Pal and P. R. Chauhan. "Environmental impact of the 73 ka Toba super-eruption in South Asia." *Palaeogeography, Palaeoclimatology, Palaeoecology* 284 (2009), 295–314.

3. What Lies Beneath

Bailey, S. "Dental morphological affinities among late Pleistocene and Recent humans." *Dental Anthropology* 14(2) (2000), 1–8.

Bar-Yosef, O., and J. Callender. "The woman from Tabun: Garrod's doubts in historical perspective." *Journal of Human Evolution* 37 (1999), 879–85.

Dean, M. C. "Tooth microstructure tracks the pace of human life-history evolution." *Proceedings of the Royal Society* B 273 (2006), 2799–808.

Dean, C., M. Leakey, D. Reid, F. Schrenk, G. Schwartz, C. Stringer and A. Walker. "Growth processes in teeth distinguish modern humans from *Homo erectus* and earlier hominins." *Nature* 414 (2001), 628–31.

Drake, N. A., R. M. Blench, S. J. Armitage, C. S. Bristow and K. H. White. "Ancient watercourses and biogeography of the Sahara explain the peopling of the desert." *Proceedings of the National Academy of Sciences USA* 108 (2011), 458–62.

Gibbons, A. "Palaeontologists get X-ray vision." *Science* 318 (2007), 1546–47.

Harvati, K., S. R. Frost and K. P. McNulty. "Neanderthal taxonomy reconsidered: Implications of 3D primate models of intra- and interspecific differences." *Proceedings of the National Academy of Sciences USA* 101 (2004), 1147–52.

Hillson, S., S. Parfitt, S. Bello, M. Roberts and C. Stringer. "Two hominin incisor teeth from the Middle Pleistocene site of Boxgrove, Sussex, England." *Journal of Human Evolution* 59 (2010), 493–503.

Irish, J. D., and D. Guatelli-Steinberg. "Ancient teeth and modern human origins: An expanded comparison of African Plio-Pleistocene and recent world dental samples." *Journal of Human Evolution* 45 (2003), 113–44.

Martinón-Torres, M., J. M. Bermúdez de Castro, A. Gómez-Robles, J.-L. Arsuaga, E. Carbonell, D. Lordkipanidze, G. Manzi and A. Margvelashvili. "Dental evidence on the hominin dispersals during the Pleistocene." *Proceedings of the National Academy of Sciences USA* 104 (2007), 13279–82.

Müller, W., H. Fricke, A. N. Halliday, M. T. McCulloch and J. A. Wartho. "Origin and migration of the Alpine Iceman." *Science* 302(5646) (2003), 862–66.

Ponce de León, M. S., L. Golovanova, V. Doronichev, G. Romanova, T. Akazawa, O. Kondo, H. Ishida and C. P. Zollikofer. "Neanderthal brain size at birth provides insights into the evolution of human life history." *Proceedings of the National Academy of Sciences USA* 105 (2008), 13764–68.

Ponce de León, M. S., and C. P. E. Zollikofer. "Neanderthal cranial ontogeny and its impli-cations for late hominid diversity." *Nature* 412 (2001), 534–38.

Ponce de León, M. S., C. P. E. Zollikofer, R. Martin and C. Stringer. "Investigation of Neanderthal morphology with computer-assisted methods." In C. Stringer, R. N. Bar-ton and C. Finlayson, eds., *Neanderthals on the Edge: 150th Anniversary Conference of the Forbes' Quarry Discovery, Gibraltar*, pp. 237–48. Oxbow Books, Oxford, 2000.

Richards, M. P., and E. Trinkaus. "Isotopic evidence for the diets of European Neander-thals and early modern humans." *Proceedings of the National Academy of Sciences USA* 106 (2009), 16034–39.

Smith, T. M., P. Tafforeau, D. J. Reid, R. Grün, S. Eggins, M. Boutakiout and J.-J. Hublin. "Earliest evidence of modern human life history in north African early *Homo sapi-ens*." *Proceedings of the National Academy of Sciences USA* 104 (2007), 6128–33.

Smith, T. M., P. Tafforeau, D. J. Reid, J. Pouech, V. Lazzari, J. Zermeno, D. Guatelli-Steinberg, A. Olejniczak, A. Hoffmann, J. Radovčić, M. Makaremi, M. Toussaint, C. Stringer and J.-J. Hublin. "Dental evidence for ontogenetic differences between modern humans and Neanderthals." *Proceedings of the National Academy of Sciences USA* 108 (2011), 8720–24.

Smith, T. M., M. Toussaint, D. J. Reid, A. J. Olejniczak and J.-J. Hublin. "Rapid dental development in a Middle Paleolithic Belgian Neanderthal." *Proceedings of the National Academy of Sciences USA* 104 (2007), 20220–25.

Spoor, F., J.-J. Hublin, M. Braun and F. Zonneveld. "The bony labyrinth of Neanderthals." *Journal of Human Evolution* 44 (2003), 141–65.

Stevens, R. E., R. Jacobi, M. Street, M. Germonpré, N. J. Conard, S. C. Münzel and R. E. M. Hedges. "Nitrogen isotope analyses of reindeer (*Rangifer tarandus*), 45,000 BP to 900 BP: Palaeoenvironmental reconstructions." *Palaeogeography, Palaeoclimatology, Palaeoecology* 262 (1–2) (2008), 32–45.

Stringer, C. B., M. C. Dean and R. Martin. "A comparative study of cranial and dental development in a recent British sample and Neanderthals." In C. J. DeRousseau, ed., *Primate Life History and Evolution*, pp. 115–52. Liss, New York, 1990.

Stringer, C. B., L. Humphrey and T. Compton. "Cladistic analysis of dental traits in recent humans using a fossil outgroup." *Journal of Human Evolution* 32 (1997), 389–402.

Stringer, C. B., E. Trinkaus, M. Roberts, S. Parfitt and R. Macphail. "The Middle Pleisto-cene human tibia from Boxgrove." *Journal of Human Evolution* 34 (1998), 509–47.

Turner, C. "Microevolution of East Asian and European populations: A dental perspec-tive." In T. Akazawa, K. Aoki and T. Kimura, eds., *The Evolution and Dispersal of Modern Humans in Asia*, pp. 415–38. Hokusen-Sha, Tokyo, 1992.

Walker, M. J., J. Ortega, K. Parmová, M. V. López and E. Trinkaus. "Morphology, body proportions, and postcranial hypertrophy of a female Neandertal from the Sima de las Palomas, southeastern Spain." *Proceedings of the National Academy of Sciences USA* 108 (2011), 10087–91.

Weaver, T. D., and J.-J Hublin. "Neandertal birth canal shape and the evolution of human childbirth." *Proceedings of the National Academy of Sciences USA* 106 (2009), 8151–56.

4. Finding the Way Forward

Ardrey, R. *African Genesis*. Atheneum, New York, 1961.

Balter, M. "Was North Africa the launch pad for modern human migrations?" *Science* 331 (2011), 20–23.

Bergman, C., and C. B. Stringer. "Fifty years after: Egbert, an early Upper Palaeolithic juvenile from Ksar Akil, Lebanon." *Paléorient* 15 (1990), 99–112.

Bermúdez de Castro, J. M., M. Martinon-Torres, E. Carbonell, S. Sarmiento, A. Rosas, J. van der Made and M. Lozano. "The Atapuerca sites and their contribution to the knowledge of human evolution in Europe." *Evolutionary Anthropology* 13 (2004), 11–24.

Bickart, K. C., C. I. Wright, R. J. Dautoff, B. C. Dickerson and L. Feldman Barrett. "Amygdala volume and social network size in humans." *Nature Neuroscience* 468 (2010) (doi: 10.1038/nn.2724).

Bigelow, R. *The Dawn Warriors: Man's Evolution toward Peace*. Little, Brown, Boston, 1969.

Brown, P., T. Sutikna, M. J. Morwood, R. P. Soejeno, E. Jatmiko, W. Saptomo and R. A. Due. "A new small-bodied hominin from the Late Pleistocene of Flores, Indonesia." *Nature* 431 (2004), 1055–61.

Crevecoeur, I., H. Rougier, F. Grine and A. Froment. "Modern human cranial diversity in the Late Pleistocene of Africa and Eurasia: Evidence from Nazlet Khater, Peştera cu Oase and Hofmeyr." *American Journal of Physical Anthropology* 140 (2009), 347–58.

Day, M. H., and C. B. Stringer. "Les restes crâniens d'Omo-Kibish et leur classification à l'intérieur du genre *Homo*." *L'Anthropologie* 94 (1991), 573–94.

Dennell, R., and W. Roebroeks. "An Asian perspective on early human dispersal from Africa." *Nature* 438 (2005), 1099–104.

Fedele, F. G., B. Giaccio and I. Hajdas. "Timescales and cultural process at 40,000 BP in the light of the Campanian Ignimbrite eruption, Western Eurasia." *Journal of Human Evolution* 55 (2008), 834–57.

Golovanova, L. V., V. B. Doronichev, N. E. Cleghorn, M. A. Koulkova, T. V. Sapelko and M. S. Shackley. "Significance of ecological factors in the Middle to Upper Paleolithic transition." *Current Anthropology* 51(5) (2010), 655–91.

Harvati, K., and J.-J. Hublin. "Morphological continuity of the face in the late Middle and Upper Pleistocene hominins from northwestern Africa—a 3D geometric morphometric analysis." In J.-J. Hublin and S. McPherron, eds., *Modern Origins: A North African Perspective*. Springer, Dordrecht, in press.

Holliday, T. W. "Body proportions in late Pleistocene Europe and modern human origins." *Journal of Human Evolution* 32 (1997), 423–47.

Holt, B. M., and V. Formicola. "Hunters of the Ice Age: The biology of Upper Palaeolithic people." *Yearbook of Physical Anthropology* 51 (2008), 70–99.

Jöris, O., and D. S. Adler. "Setting the record straight: Toward a systematic chronological understanding of the Middle to Upper Palaeolithic boundary in Eurasia." *Journal of Human Evolution* 55 (2008), 761–3.

Jöris, O., and M. Street. "At the end of the 14C time scale—the Middle to Upper Palaeolithic record of Western Eurasia." *Journal of Human Evolution* 55 (2008), 782–802.

Kuhn, S. L., et al. "The early Upper Paleolithic occupations at Ucagizli Cave (Hatay, Turkey)." *Journal of Human Evolution* 56 (2009), 87–113.

Mellars, P., and C. B. Stringer. "Introduction." In P. Mellars and C. Stringer, eds., *The Human Revolution: Behavioural and Biological Perspectives in the Origins of Modern Humans*, pp. 1–14. Edinburgh University Press, Edinburgh, 1989.

Morwood, M. J., and W. L. Jungers. "Conclusions: Implications of the Liang Bua excavations for hominin evolution and biogeography." *Journal of Human Evolution* 57 (2009), 640–48.

Mounier, A., F. Marchal and S. Condemi. "Is *Homo heidelbergens* a distinct species? New insight on the Mauer mandible." *Journal of Human Evolution* 56 (2009), 219–46.

Ramirez Rozzi, F. V., F. d'Errico, M. Vanhaeren, P. M. Grootes, B. Kerautret and V. Dujardin. "Cutmarked human remains bearing Neandertal features and modern human remains associated with the Aurignacian at Les Rois." *Journal of Anthropological Sciences* 87 (2009), 1–30.

Rightmire, G. P., D. Lordkipanidze and A. Vekua. "Anatomical descriptions, comparative studies and evolutionary significance of the hominin skulls from Dmanisi, Republic of Georgia." *Journal of Human Evolution* 50 (2006), 115–41.

Rosas, A., C. Martínez-Maza, M. Bastir, A. Garcia-Tbernero, C. Lalueza-Fox, R. Huguet, J. E. Ortiz et al. "Paleobiology and comparative morphology of a late Neandertal sample from El Sidrón, Asturias, Spain." *Proceedings of the National Academy of Sciences USA* 103 (2006), 19266–71.

Ruff, C. B. "Morphological adaptation to climate in modern and fossil hominids." *Yearbook of Physical Anthropology* 37 (1994), 65–107.

Shang, H., H. Tong, S. Zhang, F. Chen and E. Trinkaus. "An early modern human from Tianyuan Cave, Zhoukoudian, China." *Proceedings of the National Academy of Sciences USA* 104 (2007), 6573–78.

Stringer, C. B. "1970–1990: Two revolutionary decades." In K. Boyle, C. Gamble and O. Bar-Yosef, eds., *The Upper Palaeolithic Revolution in Global Perspective: Papers in Honour of Sir Paul Mellars*, pp. 35–44. McDonald Institute for Archaeological Research, Cambridge, 2010.

———. "Out of Africa—a personal history." In M. H. Nitecki and D. V. Nitecki, eds., *Origins of Anatomically Modern Humans*, pp. 151–72. Plenum Press, New York, 1994.

———. "Population relationships of later Pleistocene hominids: A multivariate study of available crania." *Journal of Archaeological Sciences* 1 (1974), 317–42.

Stringer, C. B., and C. Gamble. *In Search of the Neanderthals*. Thames & Hudson, London, 1993.

Tostevin, G. B. "Social intimacy, artefact visibility and acculturation models of Neanderthal–modern human interaction." In P. Mellars, K. Boyle, O. Bar-Yosef and C. Stringer, eds., *Rethinking the Human Revolution*, pp. 341–58. McDonald Institute for Archaeological Research, Cambridge, 2007.

Trinkaus, E., and H. Shang. "Anatomical evidence for the antiquity of human footwear: Tianyuan and Sunghir." *Journal of Archaeological Science* 35 (2008), 1928–33.

White, T. D. "Once were cannibals." *Scientific American* 265 (2001), 47–55.

White, T. D., B. Asfaw, D. Degusta, W. H. Gilbert, G. D. Richards, G. Suwa and F. C. Howell. "Pleistocene *Homo sapiens* from Middle Awash, Ethiopia." *Nature* 423 (2003), 742–47.

Wolpoff, M. H., A. ApSimon, C. B. Stringer, R. Jacobi and R. Kruszynski. "Allez Neanderthal." *Nature* 289 (1981), 823–24.

Zilhao, J., E. Trinkaus, S. Constantin, S. Milota, M. Gherase, L. Sarcina, A. Danciu, H. Rougier, J. Quilès and R. Rodrigo. "The Peştera cu Oase people, Europe's earliest modern humans." In P. Mellars, K. Boyle, O. Bar-Yosef and C. Stringer, eds., *Rethinking the Human Revolution*, pp. 249–63. McDonald Institute for Archaeological Research, Cambridge, 2007.

5. Behaving in a Modern Way: Mind Reading and Symbols

Akazawa, T., and S. Muhehen, eds., *Neanderthal Burials: Excavations of the Dederiyeh Cave, Afrin, Syria*. International Research Centre for Japanese Studies, Kyoto, 2002.

Bahn, P. G., and Jean Vertut. *Journey Through the Ice Age*. University of California Press, Berkeley, 1997.

Balme, J., and K. Morse. "Shell beads and social behaviour in Pleistocene Australia." *Antiquity* 80 (2006), 799–811.

Barham, L. "Modern is as modern does? Technological trends and thresholds in the south-central African record." In P. Mellars, K. Boyle, O. Bar-Yosef and C. Stringer, eds., *Rethinking the Human Revolution*, pp. 165–76. McDonald Institute for Archaeological Research, Cambridge, 2007.

———. "Systematic pigment use in the Middle Pleistocene of south-central Africa." *Current Anthropology* 31 (2002), 181–90.

———, ed. *The Middle Stone Age of Zambia, South Central Africa*. Western Academic and Specialist Press Limited, Bristol, 2000.

Bar-Yosef Mayer, O., B. Vandermeersch and O. Bar-Yosef. "Shells and ochre in Middle Paleolithic Qafzeh Cave, Israel: Indications for modern behavior." *Journal of Human Evolution* 56 (2009), 307–14.

Bouzouggar, A., N. Barton, M. Vanhaeren, F. d'Errico, S. Collcutt, T. Higham, E. Hodge, S. Parfitt, E. Rhodes, J.-L. Schwenninger, C. Stringer, E. Turner, S. Ward, A. Moutmir and A. Stambouli. "82,000-year-old shell beads from North Africa and implications for the origins of modern human behavior." *Proceedings of the National Academy of Sciences USA* 104 (2007), 9964–69.

Byrne, R. W. *The Thinking Ape: Evolutionary Origins of Intelligence.* Oxford University Press, Oxford, 1995.

Byrne, R. W., and L. A. Bates. "Primate social cognition: Uniquely primate, uniquely social, or just unique?" *Neuron* 65 (2010), 815–30.

Clottes, J. *Return to Chauvet Cave: Excavating the Birthplace of Art. The First Full Report.* Thames & Hudson, London, 2003.

Conard, N. J. "Cultural evolution in Africa and Eurasia during the Middle and Late Pleistocene." In W. Henke and I. Tattersall, eds., *Handbook of Paleoanthropology*, pp. 2001–37. Springer, Berlin, 2007.

——. "Cultural modernity: Consensus or conundrum?" *Proceedings of the National Academy of Sciences USA* 107 (2010), 7621–22.

——. "A female figurine from the basal Aurignacian of Hohle Fels Cave in southwestern Germany." *Nature* 459 (2009), 248–52.

Conard, N. J., M. Malina and S. Münzel. "New flutes document the earliest musical tradition in southwestern Germany." *Nature* 460 (2009), 737–40.

Coulson, S., S. Staurset and N. Walker. "Ritualized behavior in the Middle Stone Age: Evidence from Rhino Cave, Tsodilo Hills, Botswana." *PaleoAnthropology* (2011), 18–61.

Culotta, E. "On the origin of religion." *Science* 326 (2009), 784–87.

Dennell, R. "The world's oldest spears." *Nature* 385 (1997), 767–68.

d'Errico, F., H. Salomon, C. Vignaud and C. Stringer. "Pigments from the Middle Palaeolithic levels of Es-Skhul (Mount Carmel, Israel)." *Journal of Archaeological Science* 37(12) (2010), 3099–110.

Dunbar, R. I. M. "The social brain and the cultural explosion of the human revolution." In P. Mellars, K. Boyle, O. Bar-Yosef and C. Stringer, eds., *Rethinking the Human Revolution*, pp. 91–98. McDonald Institute for Archaeological Research, Cambridge, 2007.

——. "The social brain: Mind, language, and society in evolutionary perspective." *Annual Review of Anthropology* 32 (2003), 163–81.

——. "Why are humans not just great apes?" In C. Pasternak, ed., *What Makes Us Human*, pp. 37–48. Oneworld Publications, Oxford, 2007.

Gamble, C. *Origins and Revolutions: Human Identity in Earliest Prehistory.* Cambridge University Press, Cambridge, 2007.

Grün, R., C. Stringer, F. McDermott, R. Nathan, N. Porat, S. Robertson, L. Taylor, G. Mortimer, S. Eggins and M. McCulloch. "U-series and ESR analyses of bones and teeth relating to the human burials from Skhul." *Journal of Human Evolution* 49 (2005), 316–34.

Harris, S., J. T. Kaplan, A. Curiel, S. Y. Bookheimer, M. Iacoboni and M. S. Cohen. "The neural correlates of religious and nonreligious belief." *PLoS ONE* 4 (2009), e0007272.

Henrich, J. "The evolution of costly displays, cooperation and religion: Credibility enhancing displays and their implications for cultural evolution." *Evolution and Human Behavior* 30 (2009), 244–60.

Henshilwood, C. S. "The 'Upper Palaeolithic' of southern Africa: The Still Bay and Howiesons Poort techno-traditions." In S. Reynolds and A. Gallagher, eds., *African Genesis: Perspectives on Hominid Evolution*, pp. 38–50. Wits University Press, Johannesburg, 2009.

Henshilwood, C. S., and F. d'Errico, eds., *Homo Symbolicus: The Origins of Language, Symbolism and Belief.* University of Bergen Press, Bergen, in press.

Henshilwood, C. S., F. d'Errico, M. Vanhaeren, K. van Niekerk and Z. Jacobs. "Middle Stone Age shell beads from South Africa." *Science* 304 (2004), 403.

Henshilwood, C. S., F. d'Errico and I. Watts. "Engraved ochres from the Middle Stone Age levels at Blombos Cave, South Africa." *Journal of Human Evolution* 57 (2009), 27–47.

Henshilwood, C. S., F. d'Errico, R. Yates, Z. Jacobs, C. Tribolo, G. A. T. Duller, N. Mercier, J. Sealy, H. Valladas, I. Watts and A. G. Wintle. "Emergence of modern human behaviour: Middle Stone Age engravings from South Africa." *Science* 295 (2002), 1278–80.

Henshilwood, C. S., and C. W. Marean. "The origin of modern human behavior: Critique of the models and their test implications." *Current Anthropology* 44(5) (2003), 627–52.

Hovers, E., S. Ilani, O. Bar-Yosef and B. Vandermeersch. "An early case of color symbolism. Ochre use by modern humans in Qafzeh Cave." *Current Anthropology* 44 (2003), 492–522.

Hovers, E., Y. Rak and W. H. Kimbel. "Neanderthals of the Levant." *Archaeology* 49 (1996), 49–50.

Hublin, J. J. "Climatic changes, paleogeography, and the evolution of the Neandertals." In T. Akazawa, K. Aoki and O. Bar-Yosef, eds., *Neanderthals and Modern Humans in Western Asia*, pp. 295–310. Plenum, New York, 1998.

———. "The prehistory of compassion." *Proceedings of the National Academy of Sciences USA* 106 (2009), 6429–30.

Jerardino, A., and C. W. Marean. "Shellfish gathering, marine palaeoecology and modern human behavior: Perspectives from cave PP13b, Pinnacle Point, South Africa." *Journal of Human Evolution* 59(3–4) (2010), 412–24.

Klein, R. G. "Out of Africa and the evolution of human behavior." *Evolutionary Anthropology* 17 (2008), 267–81.

Kuhn, S. L., and M. C. Stiner. "Body ornamentation as information technology: Towards an understanding of the significance of early beads." In P. Mellars, K. Boyle, O. Bar-Yosef and C. Stringer, eds., *Rethinking the Human Revolution*, pp. 45–54. McDonald Institute for Archaeological Research, Cambridge, 2007.

Lewis-Williams, D. *The Mind in the Cave: Consciousness and the Origins of Art*. Thames & Hudson, London, 2002.

Lycett, S. J., M. Collard and W. C. McGrew. "Phylogenetic analyses of behavior support existence of culture among wild chimpanzees." *Proceedings of the National Academy of Sciences USA* 104 (2007), 45, 17588–92.

Marean, C. W., M. Bar-Matthews, J. Bernatchez, J. Fisher, P. Goldberg, A. Herries, Z. Jacobs, A. Jerardino, P. Karkanas, T. Minichillo, P. J. Nilssen, E. Thompson, I. Watts and H. M. Williams. "Early human use of marine resources and pigment in South Africa during the Middle Pleistocene." *Nature* 449 (2007), 905–8.

McBrearty, S. "Down with the revolution." In P. Mellars, K. Boyle, O. Bar-Yosef and C. Stringer, eds., *Rethinking the Human Revolution*, pp. 133–52. McDonald Institute for Archaeological Research, Cambridge, 2007.

McBrearty, S., and A. Brooks. "The revolution that wasn't: A new interpretation of the origin of modern human behavior." *Journal of Human Evolution* 39 (2000), 453–563.

McBrearty, S., and C. Stringer. "The coast in colour." *Nature* 449 (2007), 793–94.

Mellars, P. A. "The impossible coincidence: A single-species model for the origins of modern human behavior in Europe." *Evolutionary Anthropology* 14 (2005), 167–82.

———. "Major issues in the emergence of modern humans." *Current Anthropology* 30 (1989), 349–85.

———. "Why did modern human populations disperse from Africa ca. 60,000 years ago? A new model." *Proceedings of the National Academy of Sciences USA* 103 (2006), 9381–86.

Mellars, P., K. Boyle, O. Bar-Yosef and C. Stringer, eds. *Rethinking the Human Revolution*. McDonald Institute for Archaeological Research, Cambridge, 2007.

Mellars, P. A., and C. B. Stringer. "Introduction." In P. A. Mellars and C. B. Stringer, eds., *The Human Revolution: Behavioural and Biological Perspectives in the Origins of Modern Humans*, pp. 1–14. Edinburgh University Press, Edinburgh, 1989.

Morris, D. *The Naked Ape*. Jonathan Cape, London, 1967.

Morriss-Kay, G. M. "The evolution of human artistic creativity." *Journal of Anatomy* 216 (2010), 158–76.

Mourre, V., P. Villa and C. S. Henshilwood. "Early use of pressure flaking on lithic artifacts at Blombos Cave, South Africa." *Science* 330 (2011), 659–62.

Nowell, A. "Defining behavioral modernity in the context of Neandertal and anatomically
modern human populations." *Annual Review of Anthropology* 39 (2010), 437–52.

Pettitt, P. "The living as symbols, the dead as symbols: Problematising the scale and pace
of hominin symbolic evolution." In C. Henshilwood and F. d'Errico, eds., *Homo Symbolicus: The Origins of Language, Symbolism and Belief.* University of Bergen Press,
Bergen, in press.

Power, C. "Society as congregation—religion as binding spectacle." *Radical Anthropology*
1 (2007), 17–25.

Rizzolatti, G., M. Fabbri-Destro and L. Cattaneo. "Mirror neurons and their clinical rele-
vance." *Nature Clinical Practice Neurology* 5(1) (2009), 24–34.

Roebroeks, W., J.-J. Hublin and K. MacDonald. "Continuities and discontinuities in
Neandertal presence—a closer look at Northwestern Europe." In N. Ashton, S. Lewis
and C. B. Stringer, eds., *The Ancient Human Occupation of Britain*, pp. 113–23. Else-
vier, Amsterdam, 2011.

Stringer, C. B. *Homo britannicus.* Allen Lane, London, 2006.

Stringer, C. B., E. Trinkaus, M. Roberts, S. Parfitt and R. Macphail. "The Middle Pleistocene human tibia from Boxgrove." *Journal of Human Evolution* 34 (1998), 509–47.

Svoboda, J. "The Upper Paleolithic burial sites at Předmostí: Ritual and taphonomy." *Journal of Human Evolution* 54 (2008), 15–33.

Thieme, H. "Lower Palaeolithic hunting spears from Germany." *Nature* 385 (1997), 807–10.

Valladas, H., J.-L. Joron, G. Valadas, B. Arensburg, O. Bar-Yosef, A. Belfer-Cohen, P. Goldberg
et al. "Thermoluminescence dates for the Neanderthal burial site at Kebara in Israel."
Nature 330 (1987), 159–60.

Vanhaeren, M., and F. d'Errico. "Aurignacian ethnolinguistic-geography of Europe revealed
by personal ornaments." *Journal of Archaeological Science* 33 (2006), 1105–28.

Vanhaeren, M., F. d'Errico, C. Stringer, S. James, J. Todd and H. Mienis. "Middle Paleolithic shell beads in Israel and Algeria." *Science* 312 (2006), 1785–88.

Villa, P. "On the evidence for Neanderthal burial." *Current Anthropology* 30 (1989), 325–26.

Watts, I. "Was there a human revolution?" *Radical Anthropology* 4 (2010), 16–21.

White, R. *Prehistoric Art: The Symbolic Journey of Humankind.* Harry N. Abrams, New
York, 2003.

———. "Systems of personal ornamentation in the Early Upper Palaeolithic: Method-
ological challenges and new observations." In P. Mellars, K. Boyle, O. Bar-Yosef and C.
Stringer, eds., *Rethinking the Human Revolution*, pp. 287–302. McDonald Institute for
Archaeological Research, Cambridge, 2007.

Whiten, A. "The place of 'deep social mind' in the evolution of human nature." In C.
Pasternak, ed., *What Makes Us Human*, pp. 146–63. Oneworld Publications, Oxford,
2007.

Whiten, A., R. Hinde, C. Stringer and K. Laland, eds., *Culture Evolves.* Oxford University
Press, Oxford, in press.

Wilkins, J. "Style, symboling, and interaction in Middle Stone Age societies." *Vis-à-vis:
Explorations in Anthropology* 10(1) (2010), 102–25.

Wolpert, L. "Causal belief makes us human." In C. Pasternak, ed., *What Makes Us Human*,
pp. 164–81. Oneworld Publications, Oxford, 2007.

6. Behaving in a Modern Way: Technology and Lifeways

Adavasio, J. M., O. Soffer, D. C. Hyland, J. S. Illingworth, B. Klíma and J. Svoboda. "Per-
ishable industries from Dolní Věstonice I: New insights into the nature and origin of
the Gravettian." *Archaeology, Ethnology and Anthropology of Eurasia* 2 (2001), 48–65.

Aiello, L. C., and R. I. M. Dunbar. "Neocortex size, group size, and the evolution of
language." *Current Anthropology* 34(2) (1993), 184–93.

Alexander, R. D. *The Biology of Moral Systems*. Aldine de Gruyter, New York, 1987.

Alperson-Afil, N., G. Sharon, M. Kislev, Y. Melamed, I. Zohar, S. Ashkenazi, R. Rabinovich, R. Biton, E. Werker, G. Hartman, C. Feibel and N. Goren-Inbar. "Spatial organization of hominin activities at Gesher Benot Ya'aqov, Israel." *Science* 326 (5960) (2009), 1677–80.

Aranguren, B., R. Becattini, M. Mariotti Lippi and A. Revedin. "Grinding flour in Upper Palaeolithic Europe (25,000 years bp)." *Antiquity* 81 (314) (2007), 845–55.

Berger, T. D., and E. Trinkaus. "Patterns of trauma among the Neandertals." *Journal of Archaeological Science* 22 (1995), 841–52.

Binford, L. R. "Isolating the transition to cultural adaptations: An organizational approach." In E. Trinkaus, ed., *The Emergence of Modern Humans*, pp. 18–41. Cambridge University Press, Cambridge, 1989.

Bingham, P. M. "Human evolution and human history: A complete theory." *Evolutionary Anthropology* 9(6) (2000), 248–57.

Blaffer Hrdy, S. *Mothers and Others: The Evolutionary Origins of Mutual Understanding*. Harvard University Press, Cambridge, 2009.

Blais, J. R. E., C. Scheepers, C. Schyns, P. G. and R. Caldara. "Cultural confusions show that facial expressions are not universal." *Current Biology* 19(18) (2009), 1543–48.

Bowles, S. "Did warfare among ancestral hunter-gatherers affect the evolution of human social behaviors?" *Science* 324 (2009), 1293–98.

Boyd, R., and P. J. Richerson. "Group beneficial norms spread rapidly in a structured population." *Journal of Theoretical Biology* 215 (2002), 287–96.

Brennan, M. U. "Health and disease in the Middle and Upper Paleolithic of south-western France: A bioarchaeological study." Ph.D. diss., New York University, 1991.

Brown, K. S., C. W. Marean, A. Herries, Z. Jacobs, C. Tribolo, D. Braun, D. L. Roberts, M. C. Meyer and J. Bernatchez. "Fire as an engineering tool of early modern humans." *Science* 325(5942) (2009), 859–62.

Burt, A., and R. L. Trivers. *Genes in Conflict: The Biology of Selfish Genetic Elements*. Belknap Press, Harvard, 2006.

Buss, D. M. *Evolutionary Psychology: The New Science of the Mind*. Allyn & Bacon, Boston, 1999.

———, ed., *The Handbook of Evolutionary Psychology*. Wiley, Hoboken, NJ, 2005.

Cartmill, M. "The human (r)evolution(s)." *Evolutionary Anthropology* 19 (2010), 89–91.

Caspari, R., and S.-H. Lee. "Older age becomes common late in human evolution." *Proceedings of the National Academy of Science USA* 101 (2004), 10895–900.

Chapais, B. *Primeval Kinship: How Pair-bonding Gave Birth to Human Society*. Harvard University Press, Cambridge, MA, 2008.

Chomsky, N. *Language and Mind*. Harcourt Brace Jovanovich, New York, 1968.

Churchill, S. E., R. G. Franciscus, H. A. McKean-Peraza, J. A. Daniel and B. R. Warren. "Shanidar 3 Neandertal rib puncture wound and paleolithic weaponry." *Journal of Human Evolution* 57 (2009), 163–78.

Churchill, S. E., and J. A. Rhodes. "The evolution of the human capacity for 'killing at a distance.'" In J.-J. Hublin and M. P. Richards, eds., *The Evolution of Hominin Diets: Integrating Approaches to the Study of Palaeolithic Subsistence*, pp. 201–10. Springer, Dordrecht, 2009.

Corballis, M., and T. Suddendorf. "Memory, time, and language." In C. Pasternak, ed., *What Makes Us Human?* pp. 17–36. Oneworld Publications, Oxford, 2007.

Dart, R. A. "The predatory transition from ape to man." *International Anthropological and Linguistic Review* 1 (1953), 201–17.

Dawkins, R. *The Selfish Gene*. Oxford University Press, Oxford, 1976.

d'Errico, F., and J. Zilhão. "A case for Neandertal culture." *Scientific American* 13 (2003), 34–35.

Eaton, S. B., M. Shostak and M. Konner. *The Paleolithic Prescription: A Program of Diet, Exercise and a Design for Living*. Harper & Row, New York, 1988.

Finch, C. "Evolution of the human lifespan and diseases of aging: Roles of infection, inflammation, and nutrition." *Proceedings of the National Academy of Sciences USA* 107 (supplement 1) (2010), 1718–24.

Flannery, K. "Origins and ecological effects of early domestication in Iran and the Near East." In P. J. Ucko and G. W. Dimbleby, eds., *The Domestication and Exploitation of Plants and Animals*, pp. 73–100. Aldine, Chicago, 1969.

Fodor, J. A. *LOT 2: The Language of Thought Revisited.* Oxford University Press, Oxford, 2008.

Fox, R. *Kinship and Marriage: An Anthropological Perspective.* Cambridge University Press, Cambridge, 1996.

Froehle, W., and S. E. Churchill. "Energetic competition between Neandertals and anatomically modern humans." *PalaeoAnthropology* (2009), 96–116.

Gargett, R. H. "Middle Palaeolithic burial is not a dead issue: The view from Qafzeh, Saint Césaire, Kebara, Amud, and Dederiyeh." *Journal of Human Evolution* 37(1) (1999), 27–90.

Germonpré, M., M. V. Sablin, R. E. Stevens, R. E. M. Hedges, M. Hofreiter, M. Stiller and V. R. Després. "Fossil dogs and wolves from Palaeolithic sites in Belgium, the Ukraine and Russia: Osteometry, ancient DNA and stable isotopes." *Journal of Archaeological Science* 36 (2009), 473–90.

Goodall, J. *Through a Window: 30 Years Observing the Gombe Chimpanzees.* Weidenfeld & Nicolson, London, 1990.

Gracia, A., J. L. Arsuaga, I. Martínez, C. Lorenzo, J. M. Carretero, J. M. Bermúdez de Castro and E. Carbonell. "Craniosynostosis in the Middle Pleistocene human Cranium 14 from the Sima de los Huesos, Atapuerca, Spain." *Proceedings of the National Academy of Sciences USA* 106(16) (2009), 6573–78.

Hamilton, W. D. "The evolution of social behavior." *Journal of Theoretical Biology* 1 (1964), 295–311.

Hawkes, K., and J. F. O'Connell. "How old is human longevity?" *Journal of Human Evolution* 49 (2005), 650–53.

Henry, A. G., A. S. Brooks and D. R. Piperno. "Microfossils in calculus demonstrate consumption of plants and cooked foods in Neanderthal diets (Shanidar III, Iraq; Spy I and II, Belgium)." *Proceedings of the National Academy of Sciences USA* 108 (2011), 486–91.

Higham, T., R. Jacobi, M. Julien, F. David, L. Basell, R. Wood et al. "Chronology of the Grotte du Renne (France) and implications for the context of ornaments and human remains within the Châtelperronian." *Proceedings of the National Academy of Sciences USA* 107 (2010), 20234–39.

Keith, A. *A New Theory of Human Evolution.* Watts, London, 1948.

Kelly, R. C. *Warless Societies and the Origin of War.* University of Michigan Press, Ann Arbor, 2000.

Kittler, R., M. Kaysar and M. Stoneking. "Molecular evolution of *Pediculus humanus* and the origin of clothing." *Current Biology* 13 (2003), 1414–17.

Kuhn, S. L., and M. C. Stiner. "What's a mother to do? The division of labor among Neandertals and modern humans in Eurasia." *Current Anthropology* 47 (2006), 953–80.

Kvavadze, E., O. Bar-Yosef, A. Belfer-Cohen, E. Boaretto, N. Jakeli, Z. Matskevich and T. Meshveliani. "30,000-year-old wild flax fibers." *Science* 325(5946) (2010), 1359.

Laitman, J. T., and J. S. Reidenberg. "The evolution of the human larynx: Nature's great experiment." In M. P. Fried and A. Ferlito, eds., *The Larynx*, pp. 19–38. Plural, San Diego, 2009.

Lieberman, P. "The evolution of human speech." *Current Anthropology* 48 (2007), 39–66.

Light, J. E., M. A. Toups and D. L. Reed. "What's in a name? The taxonomic status of human head and body lice." *Molecular Phylogenetics and Evolution* 47 (2008), 1203–16.

Mercader, J. "Mozambican grass seed consumption during the Middle Stone Age." *Science* 326 (2009), 1680–83.

Miklósi, A. *Dog Behaviour, Evolution, and Cognition.* Oxford Biology, Oxford, 2007.

Mithen, S. "Music and the origin of modern humans." In P. Mellars, K. Boyle, O. Bar-Yosef and C. Stringer, eds., *Rethinking the Human Revolution*, pp. 107–20. McDonald Institute for Archaeological Research, Cambridge, 2007.

———. *The Prehistory of the Mind.* Thames & Hudson, London, 1996.

Moore, J. "The evolution of reciprocal sharing." *Ethology and Sociobiology* 5 (1984), 5–14.

Pinker, S. *The Language Instinct.* Morrow, New York, 1994.

Piperno, D. R., E. Weiss, J. Holst and D. Nadel. "Processing of wild cereal grains in the Upper Palaeolithic revealed by starch grain analysis." *Nature* 430 (2004), 670–73.

Preece, R. C., J. A. J. Gowlett, S. A Parfitt, D. R. Bridgland and S. G. Lewis. "Humans in the Hoxnian: Habitat, context and fire use at Beeches Pit, West Stow, Suffolk, UK." *Journal of Quaternary Science* 21 (2006), 485–96.

Reed, D. L., J. E. Light, J. M. Allen and J. J. Kirchman. "Pair of lice lost or parasites regained: The evolutionary history of anthropoid primate lice." *BioMedCentral Biology* 5 (2007), 7.

Reed, D. L., V. S. Smith, S. L. Hammond, A. R. Rogers and D. H. Clayton. "Genetic analysis of lice supports direct contact between modern and archaic humans." *PLoS Biology* 2 (2004), 1972–82.

Revedin, A., B. Aranguren, R. Becattini, L. Longo, E. Marconi, M. Mariotti Lippi, N. Skakun, A. Sinitsyn, E. Spiridonova and J. Svoboda. "Thirty-thousand-year-old evidence of plant food processing." *Proceedings of the National Academy of Sciences USA* 107(44) (2010), 18815–19.

Richards, M. P., P. B. Pettitt, M. C. Stiner and E. Trinkaus. "Stable isotope evidence for increasing dietary breadth in the European Mid-Upper Paleolithic." *Proceedings of the National Academy of Sciences USA* 98 (2001), 6528–32.

Rossano, M. J. "Making friends, making tools, and making symbols." *Current Anthropology* 51 (2010), 89–98.

Shea, J. J. "The Origins of lithic projectile point technology: Evidence from Africa, the Levant and Europe." *Journal of Archaeological Science* 33 (2006), 823–46.

Shea, J. J., and M. L. Sisk. "Complex projectile technology and *Homo sapiens* dispersal into western Eurasia." *PalaeoAnthropology* (2010), 100–22.

Soffer, O. "Ancestral lifeways in Eurasia—the Middle and Upper Paleolithic records." In M. H. Nitecki and D. V. Nitecki, eds., *Origins of Anatomically Modern Humans*, pp. 101–19. Plenum Press, New York, 1994.

Soffer, O., J. M. Adovasio, J. S. Illingworth, H. A. Amirkhanov, N. D. Praslov and M. Street. "Palaeolithic perishables made permanent." *Antiquity* 74 (2000), 812–21.

Sommer, J. D. "The Shanidar IV 'Flower Burial': A re-evaluation of Neanderthal burial ritual." *Cambridge Archaeological Journal* 9(1) (1999), 127–29.

Stiner, M. C. "Thirty years on the 'Broad Spectrum Revolution' and paleolithic demography." *Proceedings of the National Academy of Sciences USA* 98 (2001), 6993–96.

Svoboda, J. A. "On modern human penetration to northern Eurasia: The multiple advances hypothesis." In P. Mellars, K. Boyle, O. Bar-Yosef and C. Stringer, eds., *Rethinking the Human Revolution*, pp. 329–40. McDonald Institute for Archaeological Research, Cambridge, 2007.

Taylor, T. *The Artificial Ape.* Macmillan, Basingstoke, 2010.

Teyssandier, N. "Revolution or evolution: The emergence of the Upper Paleolithic in Europe." *World Archaeology* 40(4) (2008), 493–519.

Teyssandier, N., F. Bon and J.-G. Bordes. "Within projectile range: Some thoughts on the appearance of the Aurignacian." *Journal of Anthropological Research* 66(2) (2010), 209–29.

Tomasello, M., M. Carpenter, J. Call, T. Behne and H. Moll. "Understanding and sharing intentions: The origins of cultural cognition." *Behavioral and Brain Sciences* 28 (2005), 675–735.

Tooby, J., and L. Cosmides. "Conceptual foundations of evolutionary psychology." In David M. Buss, ed., *The Handbook of Evolutionary Psychology*, pp. 5–67. Wiley, Hoboken, NJ, 2005.

Toups, M. A., A. Kitchen, J. E. Light and D. L. Reed. "Origin of clothing lice indicates early clothing use by anatomically modern humans in Africa." *Molecular Biology and Evolution* 28(1) (2011), 29–32.

Trinkaus, E. "Late Pleistocene adult mortality patterns and modern human establishment." *Proceedings of the National Academy of Sciences USA* 108 (2011), 1267–71.

——. "Neanderthal mortality patterns." *Journal of Archaeological Science* 22 (1995), 121–42.

——. *The Shanidar Neanderthals*. Academic Press, London 1983.

Trinkaus, E., and J. Svoboda, eds. *Early Modern Human Evolution in Central Europe: The People of Dolní Věstonice and Pavlov*. Oxford University Press, Oxford, 2006.

White, R. "Systems of personal ornamentation in the Early Upper Palaeolithic: Methodological challenges and new observations." In P. Mellars, K. Boyle, O. Bar-Yosef and C. Stringer, eds., *Rethinking the Human Revolution*, pp. 287–302. McDonald Institute for Archaeological Research, Cambridge, 2007.

Wilson, D. S., and E. O. Wilson. "Evolution 'for the good of the group.'" *American Scientist* 96(5) (2008), 380–89.

Wrangham, R. *Catching Fire: How Cooking Made Us Human*. Basic Books, New York, 2009.

Wrangham, R., and R. Carmody. "Human adaptation to the control of fire." *Evolutionary Anthropology* 19 (2010), 187–99.

Zilhão, J., D. E. Angelucci, E. Badal-García, F. d'Errico, F. Daniel, L. Dayet, K. Douka, T. G. Higham, M. J. Martínez-Sánchez, R. Montes-Bernárdez et al. "Symbolic use of marine shells and mineral pigments by Iberian Neanderthals." *Proceedings of the National Academy of Sciences USA* 107 (2009), 1023–28.

Zollikofer, C. P. E., M. S. Ponce de León, B. Vandermeersch and F. Lévêque. "Evidence for interpersonal violence in the St. Césaire Neanderthal." *Proceedings of the National Academy of Sciences USA* 99 (2002), 6444–48.

7. Genes and DNA

Adcock, G. J., E. S. Dennis, S. Easteal, G. A. Huttley, L. S. Jermlin, W. J. Peacock and A. Thorne. "Mitochondrial DNA sequences in ancient Australians: Implications for modern human origins." *Proceedings of the National Academy of Sciences USA* 98 (2001), 537–42.

Bae, C. J. "The Late Middle Pleistocene hominin fossil record of Eastern Asia: Synthesis and review." *Yearbook of Physical Anthropology* 53 (2010), 75–93.

Bandelt, H.-J., V. Macaulay and M. Richards, eds. *Human mitochondrial DNA and the evolution of Homo sapiens*. Springer-Verlag, Berlin/Heidelberg, 2006.

Bowler, P. *Evolution: The History of an Idea*. University of California Press, Berkeley, 2009.

Brotherton, P., J. J. Sanchez, A. Cooper and P. Endicott. "Preferential access to genetic information from endogenous hominin ancient DNA and accurate quantitative SNP-typing via SPEX." *Nucleic Acid Research* 38 (2009), 1–12.

Bustamante, C. D., and B. M. Henn. "Shadows of early migrations." *Nature* 468 (2010), 1044–45.

Caspari, R. "1918: Three perspectives on race and human variation." *American Journal of Physical Anthropology* 139(1) (2009), 5–15.

Charlesworth, B. "Fundamental concepts in genetics: Effective population size and patterns of molecular evolution and variation." *Nature Reviews Genetics* 10(3) (2009), 195–205.

Chiaroni, J., P. A. Underhill and L. L. Cavalli-Sforza. "Y chromosome diversity, human expansion, drift and cultural evolution." *Proceedings of the National Academy of Sciences USA* 106 (2009), 20174–79.

Coop, G., K. Bullaughey, F. Luca and M. Przeworski. "The timing of selection at the human FOXP2 gene." *Molecular Biology and Evolution* 25 (2008), 1257–59.

Cooper, A., A. Rambaut, V. Macaulay, E. Willerslev, A. Hansen and C. Stringer. "Human origins and ancient human DNA." *Science* 292 (2001), 1655–56.

Cruciani, F., B. Trombetta, A. Massaia, G. Destro-Bisol, D. Sellitto and R. Scozzari. "A revised root for the human Y chromosomal phylogenetic tree: The origin of patrilineal diversity in Africa." *American Journal of Human Genetics* 88 (2011), 814–18.

Currat, M., and L. Excoffier. "Strong reproductive isolation between humans and Neanderthals inferred from observed patterns of introgression." *Proceedings of the National Academy of Sciences USA* (2011) (doi: 10.1073/pnas.1107450108).

Duarte, C., J. Maurício, P. B. Pettitt, P. Souto, E. Trinkaus, H. van der Plicht and J. Zilhão. "The early Upper Paleolithic human skeleton from the Abrigo do Lagar Velho (Portugal) and modern human emergence in Iberia." *Proceedings of the National Academy of Sciences USA* 96 (1999), 7604–9.

Edgar, H. J. H. "Biohistorical approaches to 'race' in the United States: Biological distances among African Americans, European Americans, and their ancestors." *American Journal of Physical Anthropology* 139(1) (2009), 58–67.

Edgar, H. J. H., and K. L. Hunley. "Race reconciled?: How biological anthropologists view human variation." *American Journal of Physical Anthropology* 139(1) (2009), 1–4.

Endicott, P., S. Ho, M. Metspalu and C. Stringer. "Evaluating the mitochondrial timescale of human evolution." *Trends in Ecology and Evolution* 24 (2009), 515–21.

Endicott, P., S. Ho and C. Stringer. "Using genetic evidence to evaluate four palaeoanthropological hypotheses for the timing of Neanderthal and modern human origins." *Journal of Human Evolution* 59 (2010), 87–95.

Eswaran, V., H. Harpending and A. Rogers. "Genomics refutes an exclusively African origin of humans." *Journal of Human Evolution* 49 (2005), 1–18.

Gibbons, A. "A Denisovan legacy in the immune system?" *Science* 333 (2011), 1086.

———. "Who were the Denisovans?" *Science* 333 (2011), 1084–87.

Gravlee, C. C. "How race becomes biology: Embodiment of social inequality." *American Journal of Physical Anthropology* 139(1) (2009), 47–57.

Green, R. E., A. W. Briggs, J. Krause, K. Prüfer, H. A. Burbano, M. Siebauer, M. Lachmann and S. Pääbo. "The Neandertal genome and ancient DNA authenticity." *EMBO* (2009), 1–9.

Green, R. E., J. Krause, A. W. Briggs et al. "A draft sequence of the Neandertal genome." *Science* 328 (2010), 710–22.

Green, R. E., J. Krause, S. E. Ptak, A. W. Briggs, M. T. Ronan, J. F. Simons, L. Du et al. "Analysis of one million base pairs of Neanderthal DNA." *Nature* 444 (2006), 330–36.

Green, R. E., A.-S. Malaspinas, J. Krause, A. W. Briggs, P. L. F. Johnson, C. Uhler, M. Meyer, J. M. Good, T. Maricic, U. Stenzel, K. Prüfer et al. "A complete Neandertal mitochondrial genome sequence determined by high-throughput sequencing." *Cell* 134 (2008), 416–26.

Hudjashov, G., T. Kivisild, P. A. Underhill, P. Endicott, J. J. Sanchez, A. A. Lin, P. Shen et al. "Revealing the prehistoric settlement of Australia by Y chromosome and mtDNA analysis." *Proceedings of the National Academy of Sciences USA* 104 (2007), 8726–30.

Hughes, J. F., H. Skaletsky, T. Pyntikova, T. A. Graves, S. K. van Daalen, P. J. Minx, R. S. Fulton, S. D. McGrath, D. P. Locke et al. "Chimpanzee and human Y chromosomes are remarkably divergent in structure and gene content." *Nature* 463(7280) (2010), 536–39.

Hunley, K. L., M. E. Healy and J. C. Long. "The global pattern of gene identity variation reveals a history of long-range migrations, bottlenecks, and local mate exchange: Implications for biological race." *American Journal of Physical Anthropology* 139(1) (2009), 35–46.

Jablonski, N. G., and G. Chaplin. "The evolution of human skin coloration." *Journal of Human Evolution* 39 (2000), 57–106.

———. "Human skin pigmentation as an adaptation to UV radiation." *Proceedings of the National Academy of Sciences USA* 107, supplement 2 (2010), 8962–68.

Jolly, C. "A proper study for mankind: Analogies from the Papionin monkeys and their implications for human evolution." *American Journal of Physical Anthropology*, supplement 33 (2001), 177–204.

Karafet, T. M., F. L. Mendez, M. B. Meilerman, P. A. Underhill, S. L. Zegura and M. F. Hammer. "New binary polymorphisms reshape and increase resolution of the human Y chromosomal haplogroup tree." *Genome Research* 18(5) (2008), 830–38.

Krause, J., C. Lalueza-Fox, L. Orlando, W. Enard, R. E. Green, H. A. Burbano, J.-J. Hublin et al. "The derived FOXP2 variant of modern humans was shared with Neandertals." *Current Biology* 17(21), 1908–12.

Krings, M., A. Stone, R. W. Schmitz, H. Krainitzki, M. Stoneking and S. Pääbo. "Neanderthal DNA sequences and the origin of modern humans." *Cell* 90 (1997), 19–30.

Lalueza-Fox, C. "The Neanderthal Genome project and beyond." *Contributions to Science* 5(2) (2009), 169–75.

Lalueza-Fox, C., E. Gigli, M. de la Rasilla, J. Fortea, A. Rosas, J. Bertranpetit and J. Krause. "Genetic characterization of the ABO blood group in Neanderthals." *BMC Evolutionary Biology* 8(1) (2008), 342.

Lalueza-Fox, C., C. H. Römpler, D. Caramelli, C. Stäubert, G. Catalano, D. Hughes, N. Rohland et al. "A Melanocortin 1 Receptor allele suggests varying pigmentation among Neanderthals." *Science* 318 (2007), 1453–55.

Lalueza-Fox, C., A. Rosas, A. Estalrrich, E. Gigli, P. F. Campos, A. García-Tabernero, S. García-Vargas, F. Sánchez-Quinto, O. Ramírez, S. Civit, M. Bastir, R. Huguet, D. Santamaría, M. T. P. Gilbert, E. Willerslev and Marco de la Rasilla. "Genetic evidence for patrilocal mating behaviour among Neandertal groups." *Proceedings of the National Academy of Sciences USA* 108 (2011), 250–53.

Lambert, C., and S. A. Tishkoff. "Genetic structure in African populations: Implications for human demographic history." *Cold Spring Harbor Symposia on Quantitative Biology* 74 (2009), 395–402.

Lander, E. "Initial impact of the sequencing of the human genome." *Nature* 470 (2011), 187–97.

Lari, M., E. Rizzi, L. Milani, G. Corti, C. Balsamo et al. "The microcephalin ancestral allele in a Neanderthal individual." *PLoS ONE* 5(5) (2010), e10648 (doi: 10.1371/journal. pone.0010648).

Linz, B., F. Balloux, Y. Moodley, A. Manica, H. Liu, P. Roumagnac, D. Falush et al. "An African origin for the intimate association between humans and Helicobacter pylori." *Nature* 445 (2007), 915–18.

Liu, H., F. Prugnolle, A. Manica and F. Balloux. "A geographically explicit genetic model of worldwide human-settlement history." *American Journal of Human Genetics* 79 (2006), 230–37.

Liu, W., C.-Z. Jin, Y.-Q. Zhang, Y,-J. Cai, S. Xing et al. "Human remains from Zhirendong, South China, and modern human emergence in East Asia." *Proceedings of the National Academy of Sciences USA* 107 (2010), 19201–6.

Long, J. C., J. Li and M. E. Healy. "Human DNA sequences: More variation and less race." *American Journal of Physical Anthropology* 139(1) (2009), 23–34.

Martinón-Torres, M., R. Dennell and J. M. Bermúdez de Castro. "The Denisova hominin need not be an Out of Africa story." *Journal of Human Evolution* 60(2) (2011), 251–55.

Nei, M., and A. Roychoudhury. "Genetic relationship and evolution of human races." *Evolutionary Biology* 14 (1982), 1–59.

Pääbo, S., H. Poinar, D. Serre, V. Jaenicke-Despres, J. Hebler, N. Rohland, M. Kuch, J. Krause, L. Vigilant and M. Hofreiter. "Genetic analyses from ancient DNA." *Annual Review of Genetics* 38 (2004), 645–79.

Plagnol, V., and J. D. Wall. "Possible ancestral structure in human populations." *PLoS Genetics* 2 (2006), e105.

Pollard, K. S., S. R. Salama, B. King, A. D. Kern, T. Dreszer, S. Katzman, A. Siepel, J. S. Pedersen, G. Bejerano, R. Baertsch, K. R. Rosenbloom, J. Kent and D. Haussler. "Forces shaping the fastest evolving regions in the human genome." *PLoS Genetics* 2(10) (2006), e168.

Reich, D., R. E. Green, M. Kircher, J. Krause, N. Patterson, E. Y. Durand, B. Viola, A. W. Briggs, U. Stenzel et al. "Genetic history of an archaic hominin group from Denisova Cave in Siberia." *Nature* 468 (2010), 1053–60.

Relethford, J. H. "Genetic evidence and the modern human origins debate." *Heredity* 100 (2008), 555–63.

———. "Race and global patterns of phenotypic variation." *American Journal of Physical Anthropology* 139(1) (2009), 16–22.

Rosas, A., C. Martínez-Maza, M. Bastir, A. Garcia-Tbernero, C. Lalueza-Fox, R. Huguet, J. E. Ortiz et al. "Paleobiology and comparative morphology of a late Neandertal sample from El Sidrón, Asturias, Spain." *Proceedings of the National Academy of Sciences USA* 103 (2006), 15266–71.

Sarich V. M., and A. C. Wilson. "Immunological time scale for hominid evolution." *Science* 158 (1967), 1200–203.

Smith, C., A. Chamberlain, M. Riley, C. Stringer and M. Collins. "The thermal history of human fossils and the likelihood of successful DNA amplification." *Journal of Human Evolution* 45 (2003), 203–17.

Stone, R. "Signs of early *Homo sapiens* in China?" *Science* 326 (2009), 655.

Templeton, A. "Out of Africa again and again." *Nature* 416 (2002), 45–51.

Tishkoff, S. A., M. K. Gonder, B. M. Henn, H. Mortensen, A. Knight, C. Gignoux, N. Fernandopulle, G. Lema, T. B. Nyambo, U. Ramakrishnan, F. A. Reed and J. L. Mountain. "History of click-speaking populations of Africa inferred from mtDNA and Y chromosome genetic variation." *Molecular Biology and Evolution* 24 (2007), 2180–95.

Toups, M. A., A. Kitchen, J. E. Light and D. L. Reed. "Origin of clothing lice indicates early clothing use by anatomically modern humans in Africa." *Molecular Biology and Evolution* 28(1) (2011), 29–32.

Vargha-Khadem, F., D. G. Gadian, A. Copp and M. Mishkin. "FOXP2 and the neuroanatomy of speech and language." *Nature Reviews Neuroscience* 6 (2005), 131–38.

Wainscoat, J., A. Hill, A. Boyce, J. Flint, M. Hernandez, S. Thein, J. Old, J. Lynch, A. Falusi, D. Weatherall and J. Clegg. "Evolutionary relationships of human populations from an analysis of nuclear DNA polymorphisms." *Nature* 319 (1986), 491–93.

Wall, J. D., M. P. Cox, F. L. Mendez, A. Woerner, T. Severson and M. F. Hammer. "A novel DNA sequence database for analyzing human demographic history." *Genome Research* 18(8) (2008), 1354–61.

Weaver, T., C. Roseman and C. Stringer. "Were Neandertal and modern human cranial differences produced by natural selection or genetic drift?" *Journal of Human Evolution* 53 (2007) 135–45.

White, D., and M. Rabago-Smith. "Genotype–phenotype associations and human eye color." *Journal of Human Genetics* 56 (2011), 5–7.

Yotova V., J. F. Lefebvre, C. Moreau, E. Gbeha, K. Hovhannesyan, S. Bourgeois, S. Bédarida, L. Azevedo, A. Amorim, T. Sarkisian, P. H. Avogbe, N. Chabi, M. H. Dicko, E. S. Kou' Santa Amouzou, A. Sanni, J. Roberts-Thomson, B. Boettcher, R. J. Scott and D. Labuda. "An X-linked haplotype of Neandertal origin is present among all non-African populations." *Molecular Biology and Evolution* 28 (2011), 1957–62.

Zilhão, J., and E. Trinkaus, eds. *Portrait of the Artist as a Young Child: The Gravettian Human Skeleton from the Abrigo do Lagar Velho and Its Archaeological Context.* Portuguese Institute of Archaeology, Lisbon, 2002.

8. Making a Modern Human

Ambrose, S. H. "Coevolution of composite tool technology, constructive memory, and language: Implications for the evolution of modern human behavior." *Current Anthropology* 51(S1) (2010), 135–47.

———. "Middle and Later Stone Age settlement patterns in the central Rift Valley, Kenya:

Comparisons and contrasts." In N. Conard, ed., *Settlement Dynamics of the Middle Palaeolithic and Middle Stone Age*, pp. 21–43. Kerns Verlag, Tübingen, 2001.

———. "Paleolithic technology and human evolution." *Science* 291 (2001), 1748–53.

———. "Small things remembered: Origins of early microlithic industries in Subsaharan Africa." In R. Elston and S. Kuhn, eds., *Thinking Small: Global Perspectives on Microlithic Technologies*, pp. 9–29. Archaeological Papers of the American Anthropological Association (12), Washington, DC, 2002.

———. "A tool for all seasons." *Science* 314 (2006), 930–31.

Armitage, S. J., S. A. Jasim, A. F. Marks, A. G. Parker et al. "The southern route 'Out of Africa': Evidence for an early expansion of modern humans into the Arabian Peninsula." *Science* 331 (2011), 453–56.

Atkinson, Q. "Phonemic diversity supports a serial founder effect model of language expansion from Africa." *Science* 332 (2011), 346–49.

Baker, M. "The search for association." *Nature* 467 (2010), 1135–38.

Balter, M. "Anthropologist brings worlds together." *Science* 329 (2010), 743–45.

Barker, G., H. Barton, M. Bird, P. Daly, I. Datan, A. Dykes, L. Farr, D. Gilbertson, B. Harrisson, C. Hunt, T. Higham, L. Kealhofer, J. Krigbaum, H. Lewis, S. McLaren, V. Paz, A. Pike, P. Piper, B. Pyatt, R. Rabett, T. Reynolds, J. Rose, G. Rushworth, M. Stephens, C. Stringer, J. Thompson and C. Turney. "The 'human revolution' in lowland tropical Southeast Asia: The antiquity and behavior of anatomically modern humans at Niah Cave (Sarawak, Borneo)." *Journal of Human Evolution* 52 (2007), 243–61.

Bookstein, F., K. Schäfer, H. Prossinger, H. Seidler, M. Fieder, C. Stringer, G. W. Weber, J.-L. Arsuaga, D. E. Slice, F. J. Rohlf, W. Recheis, A. J. Mariam and L. P. Marcus. "Comparing frontal cranial profiles in archaic and modern *Homo* by morphometric analysis." *Anatomical Record* 257(6) (1999), 217–24.

Bruner, E. "Comparing endocranial form and shape differences in modern humans and Neandertal: A geometric approach." *PaleoAnthropology* (2008), 93–106.

———. "Geometric morphometrics and paleoneurology: Brain shape evolution in the genus *Homo*." *Journal of Human Evolution* 47 (2004) 279–303.

Burke, A. "Spatial abilities, cognition and the pattern of Neanderthal and modern human dispersal." *Quaternary International* (2010), 1–6 (doi: 10.1016/j.quaint.2010.10.029).

Castañeda, I. S., S. Mulitza, E. Schefuß, R. A. Lopes dos Santos, J. S. Sinninghe Damsté and S. Schouten. "Wet phases in the Sahara/Sahel region and human migration patterns in North Africa." *Proceedings of the National Academy of Sciences USA* 106 (2009), 20159–63.

Chase, B. M. "South African palaeoenvironments during marine oxygen isotope stage 4: A context for the Howiesons Poort and Still Bay industries." *Journal of Archaeological Science* 37 (2010), 1359–66.

Coolidge, F. L., and T. Wynn, eds. *The Rise of* Homo sapiens: *The Evolution of Modern Thinking.* Wiley-Blackwell, Chichester, 2009.

Cronin, H. *The Ant and the Peacock: Altruism and Sexual Selection from Darwin to Today.* Cambridge University Press, Cambridge, 1991.

———. "Getting human nature right." In J. Brockman, ed., *The New Humanist: Science at the Edge*, pp. 53–65. Barnes & Noble Books, New York, 2003.

de Beaune, S. A., F. L. Coolidge and T. Wynn, eds., *Cognitive Archaeology and Human Evolution.* Cambridge University Press, Cambridge, 2009.

Ding, Y.-C., D. L. Grady, J. M. Swanson, R. K. Moyzis et al. "Evidence of positive selection acting at the human dopamine receptor D4 gene locus." *Proceedings of the National Academy of Sciences USA* 99(1) (2002), 309–14.

Douglas, K. "Culture club: All species welcome." *New Scientist* 2787 (2010), 38–41.

Flood, J. *Archaeology of the Dreamtime: The Story of Prehistoric Australia and Its People.* Yale University Press, New Haven, 1990.

Foley, R. "The ecological conditions of speciation: A comparative approach to the origins of anatomically-modern humans." In P. Mellars and C. Stringer, eds., *The Human Revolution: Behavioural and Biological Perspectives in the Origins of Modern Humans*, pp. 298–318. Edinburgh University Press, Edinburgh, 1989.

Gunz, P., S. Neubauer, B. Maureille and J.-J. Hublin. "Brain development after birth differs between Neanderthals and modern humans." *Current Biology* 20(21) (2010), 921–2.

Haslam, M., C. Clarkson, M. Petraglia, R. Korisettar et al. "The 74 ka Toba super-eruption and southern Indian hominins: Archaeology, lithic technology and environment at Jwalapuram Locality 3." *Journal of Archaeological Science* 37 (2010), 3370–84.

Henrich, J. "Demography and cultural evolution: How adaptive cultural processes can produce maladaptive losses—the Tasmanian case." *American Antiquity* 69 (2004), 197–214.

———. "The evolution of costly displays, cooperation and religion: Credibility enhancing displays and their implications for cultural evolution." *Evolution and Human Behavior* 30 (2009), 244–60.

Henrich, J., R. Boyd and P. J. Richerson. "Five misunderstandings about cultural evolution." *Human Nature* 19 (2008), 119–37.

Henrich, J., and R. McElreath. "The evolution of cultural evolution." *Evolutionary Anthropology* 12 (2003), 123–35.

Henshilwood, C. S., F. d'Errico and I. Watts. "Engraved ochres from the Middle Stone Age levels at Blombos Cave, South Africa." *Journal of Human Evolution* 57 (2009), 27–47.

The HUGO Pan-Asian SNP Consortium. "Mapping human genetic diversity in Asia." *Science* 326 (2009), 1541–45.

Kingdon, J. *Self-made Man and His Undoing*. Simon & Schuster, London, 1993.

Klein, R. G. *The Human Career*. University of Chicago Press, Chicago, 1999.

———. "Out of Africa and the evolution of human behavior." *Evolutionary Anthropology* 17 (2008), 267–81.

Klein, R. G., G. Avery, K. Cruz-Uribe, D. Halkett, J. E. Parkington, T. Steele, P. Thomas, T. P. Volman and R. Yates. "The Ysterfontein 1 Middle Stone Age site, South Africa, and early human exploitation of coastal resources." *Proceedings of the National Academy of Sciences USA* 101 (2004), 5708–15.

Lahr, M. M. *The Evolution of Modern Human Diversity: A Study of Cranial Variation*. Cambridge University Press, Cambridge, 1996.

Lahr, M. M., and R. A. Foley. "Multiple dispersals and modern human origins." *Evolutionary Anthropology* 3 (1994), 48–60.

———. "Towards a theory of modern human origins: Geography, demography and diversity in recent human evolution." *Yearbook of Physical Anthropology* 41 (1998), 137–76.

Lieberman, D. E. "Speculations about the selective basis for modern human craniofacial form." *Evolutionary Anthropology* 17 (2008), 55–68.

Lieberman, D. E., B. M. McBratney and G. Krovitz. "The evolution and development of cranial form in *Homo sapiens*." *Proceedings of the National Academy of Sciences USA* 99 (2002), 1134–39.

Marean, C. W., M. Bar-Matthews, J. Bernatchez, J. Fisher, P. Goldberg, A. Herries, Z. Jacobs, A. Jerardino, P. Karkanas, T. Minichillo, P. J. Nilssen, E. Thompson, I. Watts and H. M. Williams. "Early human use of marine resources and pigment in South Africa during the Middle Pleistocene." *Nature* 449 (2007), 905–8.

Miller, G. *The Mating Mind: How Sexual Choice Shaped the Evolution of Human Nature*. Heinemann, London, 2000.

O'Connell, J. F., and J. Allen. "Dating the colonization of Sahul (Pleistocene Australia–New Guinea): A review of recent research." *Journal of Archaeological Science* 31 (2004), 835–53.

———. "Pre-LGM Sahul (Pleistocene Australia–New Guinea) and the archaeology of early modern humans." In P. Mellars, K. Boyle, O. Bar-Yosef and C. Stringer, eds., *Rethink-

ing the Human Revolution, pp. 395–410. McDonald Institute for Archaeological Research, Cambridge, 2007.

O'Connell, J. F., J. Allen and K. Hawkes. "Pleistocene Sahul and the origins of seafaring." In A. Anderson, J. Barrett and K. Boyle, eds., *The Global Origins and Development of Seafaring*, pp. 58–69. McDonald Institute for Archaeological Research, Cambridge, 2010.

Oppenheimer, S. "The great arc of dispersal of modern humans: Africa to Australia." *Quaternary International* 202 (2009), 2–13.

Parkington, J. E. *Shorelines, Strandlopers and Shell Middens*. Creda Communications, Cape Town, 2006.

Pearson, O. M. "Postcranial remains and the origin of modern humans." *Evolutionary Anthropology* 9 (2000), 229–47.

———. "Statistical and biological definitions of 'anatomically modern' humans: Suggestions for a unified approach to modern morphology." *Evolutionary Anthropology* 17 (2008), 38–48.

Petraglia, M. D. "Mind the gap: Factoring the Arabian Peninsula and the Indian Subcontinent into Out of Africa models." In P. Mellars, K. Boyle, O. Bar-Yosef and C. Stringer, eds., *Rethinking the Human Revolution*, pp. 383–94. McDonald Institute for Archaeological Research, Cambridge, 2007.

Petraglia, M. D., M. Haslam, D. Q. Fuller, N. Boivin and C. Clarkson. "Out of Africa: New hypotheses and evidence for the dispersal of *Homo sapiens* along the Indian Ocean rim." *Annals of Human Biology* 37 (2010), 288–311.

Petraglia, M. D., R. Korisettar, N. Boivin, C. Clarkson, P. Ditchfield, S. Jones, J. Koshy et al. "Middle Paleolithic assemblages from the Indian subcontinent before and after the Toba Super-Eruption." *Science* 317 (2007), 114–16.

Pettitt, P. B. "The living as symbols, the dead as symbols: Problematising the scale and pace of hominin symbolic evolution." In C. Henshilwood and F. d'Errico, eds., *Homo Symbolicus: The Origins of Language, Symbolism and Belief*. University of Bergen Press, Bergen, in press.

———. "The Neanderthal dead: Exploring mortuary variability in Middle Palaeolithic Eurasia." *Before Farming* 1 (2002), 1–19.

Powell, A., S. Shennan and M. Thomas. "Late Pleistocene demography and the appearance of modern human behavior." *Science* 324 (2009), 1298–1301.

Revel, M., E. Ducassou, F. E. Grousset, S. M. Bernasconi, S. Migeon, S. Revillon, J. Mascle, A. Murat, S. Zaragosi and D. Bosch. "100,000 years of African monsoon variability recorded in sediments of the Nile margin." *Quaternary Science Reviews* 29 (2010), 1342–62.

Rightmire, G. P. "*Homo* in the Middle Pleistocene: Hypodigms, variation, and species recognition." *Evolutionary Anthropology* 17 (2008), 8–21.

Rohling, E. J., Q. S. Liu, A. P. Roberts, J. D. Stanford, S. O. Rasmussen, P. L. Langen and M. Siddall. "Controls on the East Asian monsoon during the last glacial cycle, based on comparison between Hulu Cave and polar ice-core records." *Quaternary Science Reviews* 28 (27–28) (2009), 3294–302.

Rose, J. I. "New light on human prehistory in the Arabo-Persian Gulf Oasis." *Current Anthropology* 51(6) (2010), 849–83.

Rosenberg, K. R., L. Zuné and C. B. Ruff. "Body size, body proportions and encephalization in a Middle Pleistocene archaic human from northern China." *Proceedings of the National Academy of Sciences USA* 103 (2006), 3552–56.

Ryosuke, K., T. Yamaguchi, M. Takeda, O. Kondo, T. Toma, K. Haneji, T. Hanihara, H. Matsukusa, S. Kawamura, K. Maki, M. Osawa, H. Ishida and H. Oota. "A common variation in EDAR is a genetic determinant of shovel-shaped incisors." *American Journal of Human Genetics* 85(4) (2009), 528–35.

Sauer, C. "Seashore—primitive home of man?" *Proceedings of the American Philosophical Society* 106 (1962), 41–47.

Scholz, C., A. Cohen, T. Johnson, J. King, M.Talbot and E. Brown. "Scientific drilling in the Great Rift Valley: The 2005 Lake Malawi Scientific Drilling Project—an overview of the past 145,000 years of climate variability in Southern Hemisphere East Africa." *Palaeogeography, Palaeoclimatology, Palaeoecology* 303 (2011), 3–19.

Scholz, C. A., T. C. Johnson, A. S. Cohen, J. W. King, J. A. Peck, J. T. Overpeck, M. R. Talbot et al. "East African megadroughts between 135 and 75 thousand years ago and bearing on early-modern human origins." *Proceedings of the National Academy of Sciences USA* 104 (2007), 16422–27.

Shea, J. "*Homo sapiens* is as *Homo sapiens* was." *Current Anthropology* 52 (2011), 1–35.

Shennan, S. "Demography and cultural innovation: A model and its implications for the emergence of modern human culture." *Cambridge Archaeological Journal* 11 (2001), 5–16.

———. "Descent with modification and the archaeological record." *Philosophical Transactions of the Royal Society B* 366 (2011), 1070–79.

Soares, P., L. Ermini, N. Thomson, M. Mormina, T. Rito, A. Röhl, A. Salas, S. Oppenheimer, V. Macaulay and M. B. Richards. "Correcting for purifying selection: An improved human mitochondrial molecular clock." *American Journal of Human Genetics* 84 (2009), 1–20.

Soffer, O. "Ancestral lifeways in Eurasia—the Middle and Upper Paleolithic records." In M. H. Nitecki and D. V. Nitecki, eds., *Origins of Anatomically Modern Humans*, pp. 101–19. Plenum Press, New York, 1994.

Stringer, C. B. "Coasting out of Africa." *Nature* 405 (2000), 24–27.

———. "Reconstructing recent human evolution." *Philosophical Transactions of the Royal Society, London (B)* 337 (1992), 217–24.

Svoboda, J. "The Upper Paleolithic burial sites at Předmostí: Ritual and taphonomy." *Journal of Human Evolution* 54 (2008), 15–33.

Texier, J.-P., G. Porraz, J. Parkington, J.-P. Rigaud, C. Poggenpoel, C. Miller, C. Tribolo, C. Cartwright, A. Coudenneau, R. Klein, T. Steele and C. Verna. "A Howiesons Poort tradition of engraving ostrich eggshell containers dated to 60,000 years ago at Diepkloof Rock Shelter, South Africa." *Proceedings of the National Academy of Sciences USA* 107 (2010), 7621–22 (doi: 10.1073/pnas.0913047107).

Tierney, J. E., J. M. Russell, Y. S. Huang and A. S. Cohen. "Northern Hemisphere controls on tropical Southeast African climate during the last 60,000 years." *Science* 322 (2008), 252–55.

Trinkaus, E., and J. Svoboda, eds. *Early Modern Human Evolution in Central Europe: The People of Dolní Věstonice and Pavlov.* Oxford University Press, Oxford, 2006.

Utrilla, P., C. Mazo, M. C. Sopena, M. Martínez-Bea and R. Domingo. "A Palaeolithic map from 13,660 calBP: Engraved stone blocks from the Late Magdalenian in Abauntz Cave (Navarra, Spain)." *Journal of Human Evolution* 57(2) (2009), 99–111.

Verschuren, D., and J. M. Russell. "Paleolimnology of African lakes: Beyond the exploration phase." *PAGES News* 17(3) (2009), 112–14.

Watts, I. "Ochre in the Middle Stone Age of southern Africa: Ritualised display or hide preservative?" *South African Archaeological Bulletin* 57 (2002), 1–14.

———. "Red ochre, body painting and language: Interpreting the Blombos ochre." In R. Botha and C. Knight, eds., *The Cradle of Language*, pp. 62–92. Oxford University Press, Oxford, 2009.

Weaver, T., C. Roseman and C. Stringer. "Were Neandertal and modern human cranial differences produced by natural selection or genetic drift?" *Journal of Human Evolution* 53 (2007), 135–45.

White, T. D., B. Asfaw, D. Degusta, W. H. Gilbert, G. D. Richards, G. Suwa and F. C. Howell. "Pleistocene *Homo sapiens* from Middle Awash, Ethiopia." *Nature* 423 (2003), 742–47.

Wynn, T. "Archaeology and cognitive evolution." *Behavioral and Brain Sciences* 25 (2002), 389–438.

Wynn, T., and F. L. Coolidge. "Beyond symbolism and language." *Current Anthropology* 51 (2010), 5–16.

———. "Did a small but significant enhancement in working memory capacity power the evolution of modern thinking?" In P. Mellars, K. Boyle, O. Bar-Yosef and C. Stringer, eds., *Rethinking the Human Revolution*, pp. 79–90. McDonald Institute for Archaeological Research, Cambridge, 2007.

———, eds. "Working memory: Beyond language and symbolism." *Current Anthropology* 51, supplement 1 (2010).

9. The Past and Future Evolution of Our Species

Ackermann, R. R. "Phenotypic traits of primate hybrids: Recognizing admixture in the fossil record." *Evolutionary Anthropology* 19(6) (2010), 258–70.

Allsworth-Jones, P., K. Harvati and C. Stringer. "The archaeological context of the Iwo Eleru cranium from Nigeria, and preliminary results of new morphometric studies." In R. Botha and C. Knight, eds., *West African Archaeology, New Developments, New Perspectives*, pp. 29–42. British Archaeological Reports International Series S2164, 2010.

Avery, D. M. "Taphonomy of micromammals from cave deposits at Kabwe (Broken Hill) and Twin Rivers in central Zambia." *Journal of Archaeological Science* 29 (2002), 537–44.

Balter, M. "Are humans still evolving?" *Science* 309 (2005), 234–37.

Barham, L., A. Pinto Llona and C. Stringer. "Bone tools from Broken Hill (Kabwe) cave, Zambia, and their evolutionary significance." *Before Farming* 2002/2 (2002); http://www.waspress.co.uk/.

Belluz, J. "Leading geneticist Steve Jones says human evolution is over." *Times* (London), 7 October 2008.

Blum, M. G. B., and M. Jakobsson. "Deep divergences of human gene trees and models of human origins." *Molecular Biology and Evolution* (2010) (doi: 10.1093/molbev/msq265).

Boyd, R., and P. J. Richerson. "Group beneficial norms spread rapidly in a structured population." *Journal of Theoretical Biology* 215 (2002), 287–96.

Cochran, G., and H. Harpending. *The 10,000 Year Explosion: How Civilization Accelerated Human Evolution*. Basic Books, New York, 2009.

Crevecoeur, I., P. Semal, E. Cornelissen and A. S. Brooks. "The Late Stone Age human remains from Ishango (Democratic Republic of Congo): Contribution to the study of the African Late Pleistocene modern human diversity." *American Journal of Physical Anthropology* (Program of the 79th Annual Meeting of the American Association of Physical Anthropologists) 141(50) (2010), 87.

Darwin, C. Obituary. http://darwin-online.org.uk/obit.

Foley, R. A., and M. Mirazón-Lahr. "The evolution of the diversity of cultures." *Philosophical Transactions of the Royal Society B* 366 (2011), 1080–89.

Gibbons, A. "Tracing evolution's recent fingerprints." *Science* 329 (2010), 740–42.

Gluckman, P., A. Beedle and M. Hanson. *Principles of Evolutionary Medicine*. Oxford University Press, Oxford, 2009.

Gould, S. J. "The spice of life." *Leader to Leader* 15 (2000), 14–19.

Gunz, P., F. L. Bookstein, P. Mitteroeker, A. Stadlmayr, H. Seidler and G. W. Weber. "Early modern human diversity suggests subdivided population structure and a complex Out-of-Africa scenario." *Proceedings of the National Academy of Sciences USA* 106 (2009), 6094–98.

Hammer, M., A. Woerner, F. Mendez, J. Watkins and J. Wall. "Genetic evidence for archaic admixture in Africa." *Proceedings of the National Academy of Sciences USA* (in press).

Hawks, J., E. T. Wang, G. Cochran, H. C. Harpending and R. K. Moyzis. "Recent acceleration of human adaptive evolution." *Proceedings of the National Academy of Sciences USA* 104 (2007), 20753–58.

Henrich, J., R. Boyd and P. J. Richerson. "Five misunderstandings about cultural evolution." *Human Nature* 19 (2008), 119–37.

Hrdlička, A. *The Skeletal Remains of Early Man*. Smithsonian Institution, Washington,
DC, 1930.

Keinan, A., and D. Reich. "Can a sex-biased human demography account for the reduced
effective population size of chromosome X in non-Africans?" *Molecular Biology and
Evolution* 27(10) (2010), 2312–21.

Laland, K. N., J. Odling-Smee and S. Myles. "How culture shaped the human genome:
Bringing genetics and the human sciences together." *Nature Reviews/Genetics* 11
(2010), 137–48.

McAuliffe, K. "The incredible shrinking brain." *Discover Magazine*, September 2010, 54–59.

Montgomery, P. Q., H. O. L. Williams, N. Reading and C. Stringer. "An assessment of the
temporal bone lesions of the Broken Hill cranium." *Journal of Archaeological Science*
21 (1994), 331–37.

Pennisi, E. "Evolutionary medicine: Darwin applies to medical school." *Science* 324(5924)
(2009), 162–63.

Premo, L. S., and J.-J. Hublin. "Culture, population structure and low genetic diversity in
Pleistocene hominins." *Proceedings of the National Academy of Sciences USA* 106
(2009), 33–37.

Relethford, J. H. "Genetic evidence and the modern human origins debate." *Heredity* 100
(2008), 555–63.

Richerson, P. J., R. L. Bettinger and R. Boyd. "Evolution on a restless planet: Were environ-
mental variability and environmental change major drivers of human evolution?" In
F. M. Wuketits and F. J. Ayala, eds., *Handbook of Evolution*, vol. 2: *The Evolution of
Living Systems (Including Hominids)*, pp. 223–42. Wiley, Weinheim, 2005.

Richerson, P. J., R. Boyd and R. L. Bettinger. "Cultural innovations and demographic
change." *Human Biology* 81 (2009), 211–35.

Richerson, P. J., R. Boyd and J. Henrich. "Gene-culture coevolution in the age of genomics."
Proceedings of the National Academy of Sciences USA (2010) (doi: 10.1073/pnas0914631107).

Ruff, C. "Variation in human body size and shape." *Annual Review of Anthropology* 31
(2002), 211–32.

Sabeti, P. C., P. Varilly et al. "Genome-wide detection and characterization of positive
selection in human populations." *Nature* 449(7164) (2007), 913–18.

Stoneking, M. "Does culture prevent or drive human evolution?" *On the Human* (2009).
http://onthehuman.org/2009/12/does-culture-prevent-or-drive-human-evolution/.

Templeton, A. "Out of Africa again and again." *Nature* 416 (2002), 45–51.

Tishkoff, S. A., F. A. Reed, F. R. Friedlaender, C. Ehret, A. Ranciaro, A. Froment, J. B. Hirbo,
A. A. Awomoyi et al. "The genetic structure and history of Africans and African
Americans." *Science* 324 (2009), 1035–44.

Trinkaus, E. "The human tibia from Broken Hill, Kabwe, Zambia." *PaleoAnthropology*
(2009), 145–65.

———. "Modern human versus Neandertal evolutionary distinctiveness." *Current Anthropology* 47 (2006), 569–95.

Wade, N. "Adventures in very recent evolution." *New York Times*, 19 July 2010. http://
www.nytimes.com/2010/07/20/science/20adapt.html.

Ward, P. "What will become of *Homo Sapiens*?" *Scientific American* 300 (2009), 68–73.

Acknowledgments

Having worked in the field of paleoanthropology for forty years, I owe a huge debt to many people, and my network of friends and collaborators seems to get larger rather than smaller as time progresses, which is gratifying. So I am not going to attempt to name and thank everyone who has helped me in significant ways, stretching back to my family and foster family, my first teachers and supervisors, and those who welcomed me all over Europe as I began to gather data for my Ph.D. But many of my fellow researchers are identified in the book and bibliography by name, or through their ideas and influences on my thinking, and I hope I have represented their views fairly and accurately. I am certainly standing on the shoulders of giants as I grapple with reconstructing our evolutionary past, but I have also been greatly helped along the way by innumerable acts of kindness and generosity. My membership in three consortia—the completed Cambridge Stage 3 Project, the NERC-funded RESET project, and the AHOB project, funded by the Leverhulme Trust—has also greatly benefited me.

For this book I am particularly grateful to Robert Kruszynski, Rebecca Varley-Winter, and Gabrielle Delbarre for their help with the bibliography, and for illustrations my thanks go to the Natural History Museum Department of Palaeontology, Photo Unit and Images Resources, and to Silvia Bello, John Reader, Francesco d'Errico, and Nicholas Conard. I am also very grateful to the editorial and production staff at Penguin Books and Henry Holt for all their work in bringing this book to publication.

Index

Page numbers in *italics* refer to illustrations.

About the Author

Chris Stringer is the author of *The Complete World of Human Evolution, Homo britannicus*, and more than two hundred books and papers on the subject of human evolution. One of the world's foremost paleoanthropologists, he is a researcher at the Natural History Museum in London and a Fellow of the Royal Society. He has three children and lives in Sussex and London.